The **CUT**Group

The CUTGroup

Civic User Testing Group
as a New Model for UX Testing,
Digital Skills Development, and
Community Engagement in Civic Tech

DANIEL X. O'NEIL
and the Smart Chicago Collaborative

To the people of the City of Chicago and the County of Cook.
If it doesn't work for you, it doesn't work.

This is not a pipe.
—René Magritte, The Treachery of Images, 1929

I mean, we're all stuck here for a while.
Let's try to work it out.
Let's try to beat it.
Let's try and work it out.
—Rodney King, Speech on May Day, 1992

Elaborate usability tests are a waste of resources.
—Jakob Nielsen, 2000

The CUTGroup: Civic User Testing Group as a new model for UX testing, digital skills development, and community engagement in civic tech

by Daniel X. O'Neil
and the Smart Chicago Collaborative
is licensed under a Creative Commons Attribution-ShareAlike 4.0 International License.

Based on a work at http://www.cutgroupbook.org/.

Manufactured in the United States of America by the
Smart Chicago Collaborative
http://www.smartchicagocollaborative.org/
@smartchicago
info@smartchicagocollaborative.org
c/o The Chicago Community Trust
225 North Michigan Avenue
Suite 2220
Chicago, IL 60601
(773) 960-6045

Made possible by funds from the John D. and Catherine T. MacArthur Foundation and the John S. and James L. Knight Foundation.

Set in Scala and ScalaSans

Library of Congress Control Number applied for
ISBN: 978-0-9907752-4-9
2nd Edition
February 2016

Contents

1. Introduction 1
A how-to on civic tech engagement. . . 1
How it works. 1
The CUTGroup is the work of
the Smart Chicago Collaborative 2

2. Origins 5
An immediate impetus. 5
Some antecedents 7

3. Components 11
UX testing. 12
Digital skills. 13
Community engagement 13

4. Tools 17
Website 17
Wufoo 18
MailChimp 18
VISA gift cards 18
Excel 19
Patterns. 19
Devices 20
Wifi 22
Candy 22
Propaganda. 22

5. Methods 25
Recruitment 25
Design. 29
Segmenting. 33
Scouting 39
Proctoring. 39
Analysis. 40
Followup 40

6. Examples 44
FreedomPop Router 44
Go to School! 46
Chicago Health Atlas 49
EatSafe.co 51
ChicagoWorksforYou.com 54
OpenStreetMap Editor. 59
EveryBlock iPhone App. 62
Waitbot 64
Foodborne Chicago 69
Build It! Bronzeville 72
Expunge.io 74
Roll with Me 78

7. Afterword 82

Introduction

A how-to on civic tech engagement

This is the CUTGroup book, an extensive how-to on the Civic User Testing Group—a community of regular Chicago and Cook County residents who get paid to test websites and apps to help create better technology. It began with a simple idea—that civic technologists should be in communion with the people they seek to serve—and it has grown to a community of more than 1,000 people who work together to make lives better through technology.

In this book, we cover in great detail how we do UX (or user experience) testing, digital skills, and community engagement in one civic tech system. We cover the hardware and software you need, methods for tester and developer recruitment, test design, location scouting, and results analysis. We show detailed budgets, exact website configurations, complete text of recruitment emails, the raw results of every test we've conducted, and all the other nuts and bolts it takes to make a CUTGroup in your city.

How it works

Here's the call to action we use on the CUTGroup website:

> *Be a tester, get paid*
>
> *The Civic User Testing Group (CUTGroup) is a community of residents of Chicago and all of Cook County who get paid to test out websites and apps. This is how it works:*
> - *Fill out the CUTGroup profile form below to sign up to be a tester. Once you do that, we will send you a $5 VISA gift card just for signing up*
> - *Once you sign up, you will receive email notifications about new test opportunities. We will ask a few more questions that relate to the technology we are going to test and you can let us know if you are available*

- *If and when you are chosen, you will test websites and apps. Testing will happen all over Cook County in public computer centers*
- *Once you complete a test, you will receive a $20 VISA gift card*

There are technology developers who make websites, mobile apps, and other tools that are made specifically for residents. By being part of the CUTGroup you will help developers make better software that will help make other lives better. If it doesn't work for you, it doesn't work!

That simple proposition has worked for us, giving us hours of great discussion and dozens of useful insights about what people really need out of civic technology in Chicago. This book shows you how it can work for you.

The CUTGroup is the work of Smart Chicago

The CUTGroup is an initiative of the Smart Chicago Collaborative, a civic organization devoted to improving lives in Chicago through technology. We work on increasing access to the Internet, improving skills for using the Internet, and developing meaningful products from data that measurably contribute to the quality of life of residents in our region and beyond.

The CUTGroup is a central program for Smart Chicago because it cuts across our three areas of focus:

- Access: the majority of our tests are conducted in public computer centers. This allows us to celebrate and promote these important connectivity points, and it also helps draw together human networks
- Skills: we focus on "on ramps," and everyone who participates in a CUTGroup test is on one of these ramps. We deliver rudimentary digital-skills training to testers, as we're often introducing them to new technology. We train developers how to design tests and engage residents in order to gather concrete feedback for their products. And we, as the Smart Chicago team, never leave a test without new skills and knowledge

- Data: by helping improve existing websites and apps, and by encouraging the creation of more effective and popular products in our field, we help deliver on the goal of creating a strong civic technology sector of the technology industry

This unique framing—access, skills, and data led us directly to the creation of this program. This framing existed in the founding of Smart Chicago, which was first conceived in the 2007 report "The City that NETWorks: Transforming Society and Economy Through Digital Excellence." This book is our attempt to share our processes on this particular program, and we have a great desire that you will share our methods as well.

References

Smart Chicago Collaborative. (2013, February 2). Civic User Testing Group. Retrieved from http://cutgroup.smartchicagoapps.org/

Smart Chicago Collaborative. (2011, July 25). Smart Chicago Collaborative. Retrieved from http://www.smartchicagocollaborative.org/

Stasch, Julia. (2007, May). The City that NETWorks: Transforming Society and Economy Through Digital Excellence. Retrieved from http://www.cityofchicago.org/dam/city/depts/doit/supp_info/DEI/CityThat-Networks.pdf

Sonja Marziano conducting a CUTGroup test at the Chicago Public Library Clearing Branch, 6423 W. 63rd Place.

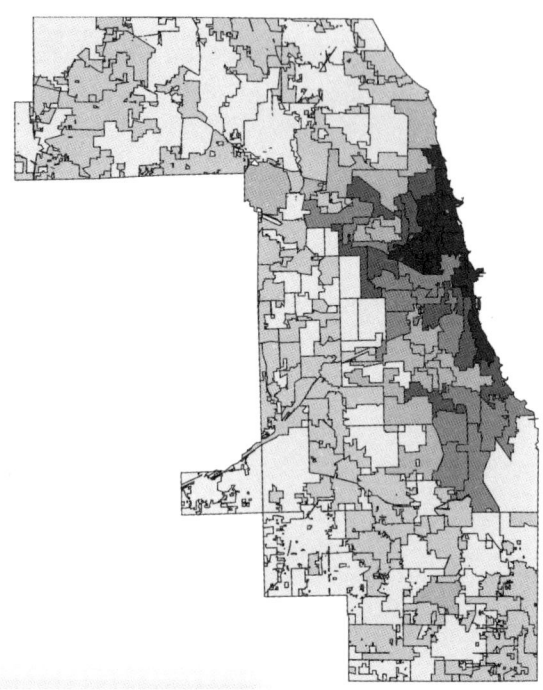

Map showing CUTGroup tester coverage across Cook County.

Origins

An immediate impetus

At Smart Chicago, all of our programs grow directly from our work. We think and learn by doing. In the fall and winter of 2012, we worked with our partner, the Illinois Science & Technology Coalition, on the Illinois Open Technology Challenge. Our mission was to bring "governments, developers and communities together in a common mission to use public data and create digital tools that will serve today's civic needs and promote economic development."

As part of our work, we did meetups up and down the state. We traveled 900 miles to conduct 8 meetups in 6 locations in 4 cities with 149 people. We worked with 12 government IT professionals to publish 138 new datasets (34 in Champaign, 15 in Rockford, 12 in Belleville, and 77 from the 42 municipalities South Suburban Mayors & Managers Association).

Here's how we described the work:

> *This is the time for anyone with an idea to present it to the group in a more complete fashion and make a pitch for people to join them. We will have lots of materials that will help you express yourselves— easels, large Post-Its, markers, etc. You'll want to talk about what data you'd like to use and what community issues you want to solve. We also ask you two questions when signing up for the Meetup—the more thinking you do before the event, the more you will get out of it.*
> - *Do you have an idea for an app that you'd like to submit? Let us know!*
> - *Do you have a community issue you'd like to address through data and technology? If so, can you describe?*

We tried to communicate the idea that everyone was welcome, whether you had "an idea for an app" (which assumes some fluency with technology) or just a "community issue you'd like to address," you could join a team and win money.

As you can see from the overall numbers, we had some success in getting people out for the meetings. We met plenty of local developers and were able to introduce them to city officials. Midway through, we realized we had to be more aggressive in outreach to community members, so we sought and received coverage on the nightly news in Rockford and afternoon radio in Champaign. Here's a snip from the news coverage:

> *With countless amounts of data, the government is seeking a way for the public to access the information to solve common problems. The City of Rockford Information Technology Department invites the community to take the Illinois Open Technology Challenge. If you develop an app to solve a problem using Rockford data, you could win $15,000. Smart Chicago Collaborative Executive Director Dan O'Neil encourages anyone with an idea to check out illinoisopentech. org. "You can make technology that's actually in the service of the people. That's the idea. That we can change Rockford, change your city and change your world with technology."*

But we failed at getting regular residents to show up at these meetings. I realized that with a value proposition that starts with "If you develop an app," there was no way we were going to get regular people to show up. We were offering $15,000 in prizes in four cities, but our program was too involved:

- Come to a meeting on a weeknight
- Develop/present an original idea for an "app"
- Persuade one or more developers to build the idea
- Follow the process through to completion

- Submit the finished site/app
- Prosper

When we got to Belleville—as far south as one could get in Illinois—we had the mayor, some developers from St. Louis, and zero members of the public. There had to be an easier way.

Some antecedents

By January 2013, I was definitely casting about for an easier way to get regular residents involved in civic tech. At the same time, I was pulling together thoughts around the limitations of civic hacking—the practice of local developers making technology with local data to serve local needs. I put those thoughts down in a blog post called "Turning Civic Hacking into Civic Innovation." Here's a snip:

> *There are lots of reasons why civic hacking works here in Chicago—a rich baseline of data and technology, an engaged developer community, real discussions with government about policy and data, and the support of institutions are all important factors.*
>
> *But what we're missing most is sustained engagement with the residents of the city of Chicago. That's how we can turn mere hacking into real innovation. The magic combination of government, developers, and community members is what we're after.*

A rich baseline of public data projects, an engaged developer community, government that cares, and support from institutions. But we had to appeal to regular residents. I turned to my experience.

- In the summer of 2003, I taught a course of 15 elementary and high school students in a weeklong "computer camp"
- In 2004, I did bilingual website training for Spanish speakers in Lincoln Square. I created and led a number of bilingual computer-training sessions for the large (but mostly invisible to the

Caucasian population at the Church) Spanish-speaking community at a parish in the North Side. I saw the need, designed tools to fill the need, and conducted the training myself

- In 2006, I developed a custom 9-hour "Websites for Small Businesses" course taught in three-hour stretches over three days. This was open to the public, and I taught all comers. Almost everyone had very low skills. Many had never had an email account before

These experiences—setting up shop, in public computer centers, for all comers—were formative for me. My experience as a co-founder of EveryBlock was also important. Everyone on the team was expected to answer user emails, which consisted of bug submissions, feature requests, and comments of all manner.

With all of this in mind—our current inability to attract community members and my past experience in working directly with people around technology—I came up with the concept of the CUTGroup.

Our motto: "If it doesn't work for you, it doesn't work."

References

Illinois Open Technology Challenge. (2012, December). Retrieved from http://illinoisopentech.org/

O'Neil, Daniel X. (2013, April 4). The Illinois Open Technology Challenge, So Far. Retrieved from http://illinoisopentech.org/the-illinois-open-technology-challenge-so-far-2/

O'Neil, Daniel X. (2013, January 18). News Coverage: ILOpenTech Rockford Meetup. Retrieved from http://illinoisopentech.org/news-coverage-ilopentech-rockford-meetup/

O'Neil, Daniel X. (2013, January 16). *Turning Civic Hacking into Civic Innovation*. Retrieved from http://www.smartchicagocollaborative.org/turning-civic-hacking-into-civic-innovation/

O'Neil, Daniel X. (2004, August 15). *Training: Bilingual (Spanish/English) Computer Training*. Retrieved from http://www.danielxoneil.com/2004/08/15/training-bilingual-spanish-english-computer-training/

O'Neil, Daniel X. (2003, August 23). *Training: Week-Long Computer Camp for Kids at Wright College*. Retrieved from http://www.danielxoneil.com/2003/08/23/training-week-long-computer-camp-for-kids-at-wright-college/

Belleville City Hall

Sixteenth Sunday in Ordinary Time — July 22, 2012

CARDENAL FRANCIS GEORGE: EL CONCILIO VATICANO II

Esta es la primera de una serie de breves reflexiones sobre el propósito y la obra del Concilio Vaticano II, escrito para los boletines parroquiales con motivo de la celebración del quincuagésimo aniversario del Concilio a partir del próximo 11 de octubre.

Estimados Hermanos y Hermanas en Cristo:

Este próximo 11 de octubre, la Iglesia marcará el quincuagésimo aniversario de la sesión inaugural del Concilio Vaticano II. Este gran Concilio fue el evento católico más significante del último siglo, y sus deliberaciones y decretos continuarán a influir a la Iglesia, al igual que los actos de los concilios ecuménicos anteriores. El credo que decimos durante Misa cada domingo, por ejemplo, viene de los primeros Concilios de la Iglesia durante el cuarto y quinto siglo. El Vaticano II fue el Concilio Ecuménico vigésimo primero de la Iglesia Católica.

¿Por qué el Papa Juan XXIII hizo un llamado al Concilio? En una carta con fecha del 25 de diciembre de 1961 (*Humanae salutis*), el Papa explicó que él esperaba que el Concilio cambiará la relación que existía entre la Iglesia y el mundo. Llamó al Concilio porque el mundo del vigésimo siglo tan estaba dividido por racismo, nacionalismo, comunismo, nazismo, guerras y odios de todo tipo que los hombres y mujeres ya habían olvidado que la raza humana es una familia humana. ¿Quién le diría al mundo que todos somos hermanos y hermanas? El Papa Juan sabía que la misión de la Iglesia Católica es presentar al mundo a su salvador en todos los tiempos, y creía que la unidad de la Iglesia debería de ser una levadura para cambiar el mundo. El Concilio Vaticano II fue llamado un concilio pastoral porque su propósito no era analizar las enseñanzas de la Iglesia directamente sino ver que podía ser cambiado en la Iglesia a fin de colocarla en el mejor diálogo con un mundo rumbo a la autodestrucción. El Concilio fue misionero en su propósito; para salvar el mundo de sí mismo, la Iglesia tendría que abrir diálogos en todos los frentes.

Con el fin de entablar un diálogo, elementos de la vida de la Iglesia eran analizados y a veces cambiados o ajustados. Los documentos principales del Concilio analizaron el culto divino de la Iglesia, las fuentes de la autorrevelación de Dios, y la naturaleza de la Iglesia misma, y la Iglesia en nuestros tiempos. Cada cuatro de estos temas serán discutidos en futuros artículos del boletín. Espero que al recordar el propósito y la enseñanza del Concilio Vaticano II renueve el sentido de misión de la arquidiócesis en el año que viene. Continuaremos a implementar el Plan Estratégico Pastoral para la Arquidiócesis y lo haremos a luz del Año de Fe Universal de la Iglesia, recientemente llamado por el Papa Benedicto XVI, programado para comenzar en el aniversario de la sesión inaugural del Concilio. Esperamos que nuestros corazones puedan abrir nuevamente al conmemorar este gran evento en la historia de la Iglesia. Que Dios los bendiga.

Sinceramente en Cristo,
Cardenal Francis George, O.M.I.
Arzobispo de Chicago

¿ALGUNA VEZ HAS PENSADO SER SACERDOTE?

Para información acerca del sacerdocio, contacte al Padre Brian Welter al bwelter@archchicago.org o al 312-534-8298. Para información acerca de la vida religiosa, contacte a la Hermana Elyse Ramirez, OP al 312-534-5240 o al eramirez@archchicago.org. Para información y para el programa del Diaconado Permanente, contacte al Diácono Bob Puhala al 847-837-4562 o al bpuhala@usml.edu.

ANIVERSARIO DE BODAS DE ORO

La Misa de Aniversario de Bodas de Oro será el domingo, 23 de septiembre del 2012 a las 2:30 p.m. in Holy Name Cathedral, 735 N. State St. Las parejas que se casaron en 1962 y están interesadas en asistir esta celebración deberán hablar a su parroquia para registrarse. Para más información, visite al www.familyministries.org o llame a la Oficina de Ministerio para la Familia al 312-534-8351.

INTENCIONES DE MISAS

Para pedir una intención especial para una misa en inglés o español en el año 2012, por favor visite la oficina parroquial o llame al 773-539-3176. La donación sugerida es de $10 por misa.

EXPOSICIÓN DEL SANTÍSIMO

Haga los viernes un buen día para orar y estar con el Señor. La iglesia estará abierta los viernes desde las 9:30 a.m. hasta las 5 p.m. Entre por la puerta que se halla sobre la Claremont, suba las escaleras y vire hacia la izquierda. Suba cinco escalones más y pase por la sacristía y detrás del altar. Allí entrará a la capilla de Veneración, donde el Santísimo estará expuesto. Traiga su Biblia, rosario, libros y sus pensamientos. Pase algún tiempo con el Santísimo. Si usted lo desea inscríbase para una hora de oración cada viernes. Favor de llamar al 773-539-3176. Pare en la capilla durante el día para pasar un rato con el Señor en oración.

¡**Bienvenidos!** ¿Es tu primera vez en nuestra parroquia? ¡Bienvenidos!
¿Se ha mudado recientemente? ¡Que Dios bendiga su nuevo hogar!
Por favor visite o contacte la oficina parroquial al 773-539-7510 para registrarse o arreglar su registración actual. ¡*Gracias!*

Queen of Angels parish communications in Spanish led to bilingual computer training.

Components

There are three components to what we do with the CUTGroup: UX testing ("user experience testing," also called "usability testing," a technique used in user-centered interaction design to evaluate a product by testing it with users); digital skills (which we define broadly as the human ability to get things done on computers); and community engagement (which, in our context, is defined as a process of building relationships for the purpose of collectively making lives better through technology).

One of the most important aspects of our work is that it does each of these three separate things pretty well, but none of them really well. The only thing we do really well is the CUTGroup itself.

We struggle at Smart Chicago with how prescriptive we should be with our program. Clearly, any kind of user testing is helpful to the technology developers. The teaching and learning of digital skills is a worthwhile act, regardless of context. And any time civic hackers can get with community members—in any setting, for any purpose—that's a good thing.

As you'll see in the Methods chapter and in the examples of tests we've conducted, different CUTGroup components take on higher or lower levels of importance, depending on the particular nuances of the need for any given project.

Sometimes, the app isn't made yet, and we're testing the relative value of a concept. In others, we're testing a mature website that has some user interface issues, and just want to get bug reports. Since we make a lot of technology, and we are engaged with audiences and experts all the time, sometimes we just want to have a structured meeting that helps us think fresh and re-engage with some of the people who matter most.

But we've settled on the belief that the melding of these three components is what makes the CUTGroup the CUTGroup—

its essence. It would therefore be impossible to say that you're running a CUTGroup program near you without being devoted to building these components in some semblance of equality and strength.

So let's take a look at each component, talk about how they fit together, and lay down some markers around what we think are the minimum elements for a viable CUTGroup.

UX testing

We have been careful to design the CUTGroup as legitimate UX testing. By that we mean the developers who work with us get specific, actionable feedback from relevantly situated users. We can obtain both quantitative and qualitative info, discovering how dozens of people answer the same question and drill into a deep conversation about any particular feature. We've found glaring bugs, discovered unique insights, and helped people plan new product releases.

We stray from UX test design principles, however, in a very key way: by requiring the developer to participate in the test. We deprecate this and other more social-science aspects of classic UX testing (never coaching the user, for instance) because it's not conducive to the other two components.

There is an odd dynamic here, however. At Smart Chicago, we have a broad mission to help build the civic innovation sector of the technology industry. By this, we mean to deliberately situate the work of civic hackers firmly in the broader market for consumer-focused technology.

By cutting against the grain of UX testing methods designed to align a software product with the needs and desires of people who might buy it, we risk keeping the sector in an immature state of development. We justify this by seeing it as a temporary condition of the field. There is currently so little engagement with regular users, and very little tradition around user testing, that we've developed this hybrid to account for that.

The minimum UX testing element that must be included in a CUTGroup program is the delivery of concrete, actionable direction to the developer that has been generated directly from testers who have been deliberately chosen for the test.

Digital skills

This is the component for which there is probably less precedence in the work of a typical civic hacking aficionado. But it's a really important one to our work at Smart Chicago, which was formed out of conversations about the digital divide.

A CUTGroup test typically consists of a mix of 10-20 people of widely varying digital skills, all learning and teaching each other, in the same room, out in plain public view. It becomes something of a salon. People get to know each other, laugh, eat candy, and pick up digital skills that they didn't expect to get when they first signed up. This goes for the developers, too—they are learning how to conduct UX tests and to make their sites better.

The more professional-grade digital-skills teachers would want to formalize this learning into a structure that allows the tester to move along a continuum of learning toward an established goal. We gave up on that idea because it would cost too much money (we can't pay people a $20 VISA gift card every time they show up for a class), and because it would defeat the informal but focused nature of a test.

So the minimum digital-skills element that must be included in a CUTGroup program is simple: a spirit of learning and a commitment that all of the testing happen in the same room, with all of the messiness that goes along with that.

Community engagement

This is the component that has surprised us the most, and the one for which we did the least amount of planning.

We always knew we wanted to conduct the tests in public computer centers and mainly in libraries. We love the vibe there and, to us, these are the most accessible and open places we could imagine. There are so many structural elements of the library—their wide geographic range, onsite wifi, community rooms, open architecture, and accessibility compliance—that our program slots in well when we center it there.

But we were surprised to see just how much people like the CUTGroup. We always ask the same last two questions on every CUTGroup test form. One is quantitative ("Did you like this CUT-Group test?") and the other is qualitative ("Anything else to add?"). Out of the hundreds of testers, one has answered "no" to the first question, and we've received dozens of positive comments about the experience.

People often have to be shepherded out of the meeting room as the lights go off in the library. We all have a really good time, and sometimes we're able to talk about difficult things (like how to get to school in an era of school closings) in very flat, constructive ways. The CUTGroup is a joy. To anyone who has ever done civic engagement (which can be a slog), you understand that this is a big deal.

From the start, we've been completely committed to drawing together as many people as we can from every part of town and all sorts of backgrounds into a single set of people on a common mission.

We differ from the practice of community organizers, however, because we don't drill down into specific geographic areas or subject matters, and we don't attempt to achieve specific social or policy outcomes. However, this leads to trust among our users because we stay focused on what we said we were interested in when they signed up: We give them money, and they give us feedback on technology.

In order to make the CUTGroup representative of the community, and to serve these larger goals around development, the minimum community engagement element that must be included in a CUTGroup program is the recruitment of a large and diverse tester base.

References

Wikipedia. (n.d.) Usability testing. Retrieved from http://en.wikipedia.org/wiki/Usability_testing

Wikipedia. (n.d.). Community engagement. Retrieved from http://en.wikipedia.org/wiki/Community_engagement

This map shows the wide geographic (and, by extension socioeconomic) range of people who came to a test of OpenStreetMap. Some people came 20 miles on a January evening when it was 10 degrees Fahrenheit with light snow.

Developer Fernando Diaz tests his food inspection site with three residents in the Kelly Library.

Tools

Following is a list of tools we use in the CUTGroup program, along with a rudimentary take on the costs for each. Our focus in pricing is to determine the bare minimum it takes to start up a reputable CUTGroup program with materials on hand.

Website

We start with a public-facing website to present the program. The website itself is just a few files hosted on GitHub.

The CUTGroup website is just three pages of information about the program, a map of participants, and embedded Wufoo signup forms for the general public and developers. This website uses Jekyll to generate static HTML pages that can be hosted in Amazon S3 for pennies a month.

CUTGroup Signups is a very small application that keeps the map of signups up-to-date. It is written in Ruby and runs on Heroku. After a CUTGroup participant completes the signup form, Wufoo sends the result to the CUTGroup Signups application. The application reads the ward of the participant and increments the number of signups for that ward. In July 2015, we added all Cook County municipalities to this map.

This is certainly not a requirement, but we think it is effective in encouraging a geographically diverse set of people to join. It's always nice to see that people you know will be at a party.

If you're not comfortable working with the files on GitHub, you can make the site any way you want, using any tools you are familiar with. WordPress is a very easy site creation tool—we highly recommend it.

If you have no way to set up a website, then just a Facebook page will do as well. The main thing here is that you have to have some sort of wrapper for the signup form—some URL that people can share with their friends and family that explains what your program is all about and displays or links to the signup form.

Costs: development time only.

Wufoo

We use Wufoo for information collection. Wufoo is a wonder. It is super easy to set up, it has powerful reporting tools, and it allows you to create sophisticated ways for sharing data among team members and between systems.

You are free to use any service you like, but given the features and pricing of Wufoo, we don't see any reason to use anything else.

Costs: $14.95 per month, given the number of forms and fields we think you need to set up a good program. Free if you are just trying it out and want to collect less information from users.

MailChimp

We use MailChimp for all of our outbound communication with testers. It has great templates, it integrates really well with other systems like Wufoo, and it allows us to segment our testers in logical ways. There are many email management systems, but we're impressed with MailChimp.

Costs: free at first. It took us a long time before we had to upgrade to an account costing $45 per month. Keep in mind that we also use MailChimp for other email marketing needs in our organization, so the costs of the tool are spread across the entire organization, not just this program.

VISA gift cards

We use Awards2Go VISA Award Card as our vendor for the incentives we use for recruiting and compensating testers. Gift cards are the central tool for tester recruitment.

Costs: $7.07 per sign-up card, excluding staff costs for managing card distribution.

Excel

From the very start, we have developed our own custom software to manage the CUTGroup. But Microsoft Excel has been an important make-ready tool for wrangling data. Excel was pre-installed on our office computers, but if you don't have it, use an alternative like LibreOffice or Google Sheets.

Costs: free

Patterns

As part of this project, we've developed software and processes that allow us to manage such a large group of people testing a wide variety of hardware and software across the city. The main tool is Patterns (which we code named "Kimball" because that's a street near Chris Gansen's house), which allows us to segment our participants by lots of criteria (location, device, Internet connection, etc.).

Patterns uses data that we gather from other web-based systems like MailChimp (for outbound email notifying CUTGroup members about testing opportunities) and Wufoo (for managing metadata about testers and their availability). Each of these systems has very strong APIs that allow us to move data in and out of Patterns. It makes operating the CUTGroup a breeze. This kind of lightweight approach is at the heart of our philosophy here at Smart Chicago.

As you can see by looking at the Patterns code repository README page on GitHub, we have a lot of work to do on this software:

- Events
  - Invite
  - RSVP

- Attendance tracking
- Reminder emails
- Programs
  - Associate results
- People
  - Add arbitrary fields
  - Attach photograph
  - Attach files
  - Link with their social networks
  - Show activity streams
  - Track program status (e.g. has received VISA card)
  - Show output from Tribune boundaries service on individual person pages
- Backend
  - Terms of service/privacy policy
  - Managed access to anonymized data for research
  - Audit trails
  - Comments on all objects

Costs: developer time.

Test Devices

We sometimes bring a number of devices to make our tests run smoothly. In other instances, it's important that the users bring their own devices. This allows us to get a better understanding of how testers use their own devices, and they feel more comfortable in the test. We also get a more diverse set of results using these devices. Bringing extra iPads and laptops is definitely a good idea, because pretty much anything can and will go wrong when you are out in the real world engaging with real people.

IPads are also a good idea to have for testers with low vision because of its robust voice command functionality and ability to easily zoom in.

We publish details on lots of the equipment we use for all of our different programs on our website. Keep tabs on our site and you can keep up with current devices, model numbers, prices, and configurations.

Costs: varied — use personal devices (free)

Recording Devices

Documenting opengov and civic engagement events is an important component of our overall mission at Smart Chicago, and it is an important component of traditional user testing. I also have a passion for photography and documentation, so it is a deep part of our culture.

Once again, the idea is to press existing devices into service. Any cellphone camera will suffice when it comes to documenting the

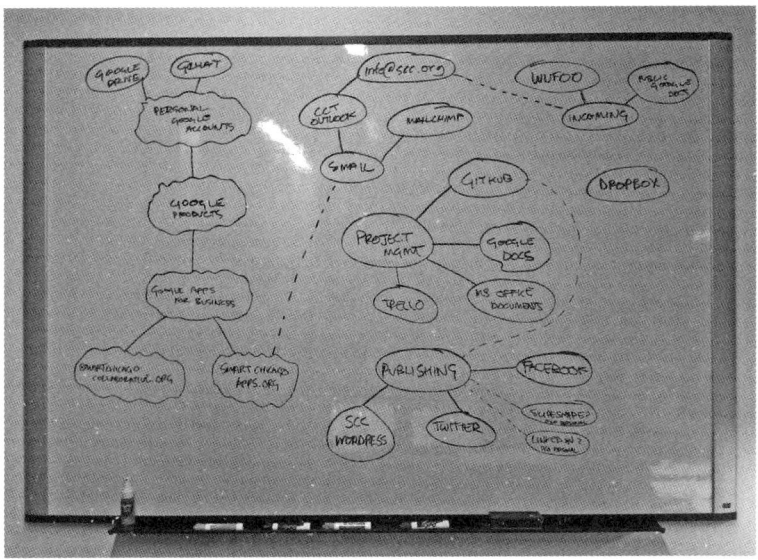

Here's a look at a schematic of Smart Chicago systems. Diagram by Chris Gansen.

tests, either by photo or video. Again, we publish specs on our own equipment to our website.

Costs: varied — use personal devices (free)

Wifi

If you're doing mobile technology work, you've got to be in control of your own connection to the Internet. Even though most of our tests are conducted in public libraries, and they all have public wifi, it is essential to carry a backup. We use Mobile Citizen.

Costs: $120 for signup, $10/month for service

Candy

Ya gotta have something to eat. Also: sharing is caring. We bring multiple types of candy (hard candy, mini chocolate bars, sour twists, etc.) Stay away from messy things (no Doritos) and, just like in elementary school, make sure you have enough for the whole class.

Propaganda

We do lots of work with computer centers, and this is an opportunity to get our message out. We leave information about CUTGroup and our Connect Chicago program and also flyers for anything else we happen to have going on.

References

CUTGroup. (n.d.). CUTGroup. Retrieved from https://github.com/smartchicago/cutgroup

Preston-Werner, Tom. (2014) Jekyl. Retrieved from http://jekyllrb.com/

Ruby. (n.d.). Retrieved from https://www.ruby-lang.org/en/

Heroku. (n.d.). Retrieved from https://www.heroku.com/

CUTGroup Signups. (n.d.). CutGroup-Signups. Retrieved from https://github.com/smartchicago/cutgroup-signups

WordPress. (n.d.). Retrieved from https://wordpress.com/

Tumblr. (n.d.). Retrieved from http://tumblr.com/

Awards2Go VISA Award Card. (n.d.). Retrieved from http://www.awards2go.net/

LibreOffice. (n.d.). Retrieved from http://www.libreoffice.org/

Wufoo. (n.d.). Retrieved from http://www.wufoo.com

MailChimp. (n.d.). Retrieved from http://www.mailchimp.com

Google Sheets. (n.d.). Retrieved from https://docs.google.com/spreadsheets/

Kimball. (n.d.). Retrieved from https://github.com/smartchicago/kimball

Smart Chicago Collaborative. (2013, October 11). Hardware and Software for CUTGroup and Civic Hacker Events. Retrieved from http://www.smartchicagocollaborative.org/hardware-and-software-for-cutgroup-and-civic-hacker-events/

Mobile Citizen (n.d.). Retrieved from http://mobilecitizen.org

Connect Chicago. (n.d.). Retrieved from http://weconnectchicago.org

The ever-present candy.

Methods

In our discussion of the components of the CUTGroup, we were careful not to be prescriptive, but we were able to lay down some markers around what we think are the minimum elements for a viable CUTGroup.

In our chapter listing the tools of the CUTGroup, we showed that it took very little money to start a program, and much of the necessary materials were laying around any civic tech operation.

As you are considering whether and how to implement a CUTGroup program near you, we urge you to consider using the suggested methods we describe below.

Recruitment

The most fundamental concept behind recruitment of testers is the idea of individual relationships. The power of the Internet is that 1:1 relationships are endlessly formed and re-formed in our experience of the web.

Our testers come from every ward in the city because we spend the time and make the effort it takes to do that. We've done flyering campaigns at all 12 City Colleges of Chicago. We analyzed our signups and did flyering campaigns at the 25 public libraries where our efforts were lagging. Then we narrowed down our efforts to just the 10 libraries in the two wards where we still couldn't get signups. We did a mass email campaign to more than 10,000 subscribers of Nightlife magazine.

In community engagement, there are lots of high-quality organizations made up of hundreds of regular residents, and the temptation is to work with those groups to sign up CUTGroup testers in one swoop. I urge you to take the long way around—recruit individual testers, on your own, to start off an independent relationship with each tester.

When we get a new test, we do work with the developer or organization to bring new testers in. The difference is once they are in the CUTGroup, they are part of our community and we build a new relationship with them.

The power of the VISA gift card

Once a resident signs up to be part of the CUTGroup, we send them a $5 VISA gift card. If and when they are chosen to test a civic app, we give them a $20 VISA gift card.

The VISA gift card is the most essential element of recruitment. There are all sorts of reasons why people might join the CUTGroup—to get involved with technology, give back to their community, or just plain get out of the house. But money is a good motivator, and respects people for their time.

We chose this incentive very deliberately. It is the most open, accessible, fungible incentive we could give—the one most akin to cash. Unlike an Amazon gift card, it is useful in the real world and can be used by walking outside and going into a corner store. We cannot assume all testers shop online or have the digital skills to use an online gift card. It's also easily transferrable.

Plus, the gift card gives us a good way to do a simple validation that the tester actually lives (or at least has access to a mailbox) in the area we're interested in—Chicago.

When purchasing gift cards for your program, there are four main considerations: type of cards, cost and fees associated per card, quantity, and expiration date.

There are different types of gift cards out there: prepaid VISA or MasterCard gift cards, store-issued, bank-issued, online, etc.

Other gift card types also include specific store- or bank-issued gift cards. Sites like ScripSmart can provide comparisons between gift cards, and they can give you an idea of what you need to ask about before purchasing your own cards.

By purchasing these cards, we are spending more than face value on fees, and have to take time to mail them out, but the value of accessible gift cards is worthwhile for the goals of our organization and this program.

That said, we have done a lot to cut our costs in managing the gift cards. When Sonja Marziano joined the team in September 2013, she cut costs from $10.17 per card to $7.07 per card. Here's a breakdown of costs with our current gift card vendor, Awards2Go VISA Award Card:

- Face value of card: $5.00
- Card processing fee: $1.75
- Credit card order fee (1% of total order): $0.05
- Shipping fee for 100: $27.00
- **Total: $7.07 per card**

Once you've got a start, build on the existing network

Last October, we learned that we still had a lot of gift cards that were about to expire and some were already expired. With the gift cards that expired, we lost out because the cost of the fees to "restock" the cards would be higher than the value of the cards we would receive.

We also had 118 $20 and 103 $5 gift cards that were going to expire at the end of November. If we sent back these cards to the vendor, we would only receive half the value of the card in return.

So we had to think up some creative ways to use the cards. The most successful campaign was "Refer a friend." Here was the pitch:

Hi <<First Name>>,

We have some $5 VISA gift cards that are about to expire at the end of the month and we need your help! We want to make sure these gift cards are used, so if you refer a friend to join the Civic User Testing Group (CUTGroup) by the end of the day this Thursday, we will send you another $5 VISA gift card! Forward this email to your friends and have them complete this form (link).

How it works:
- We will send you a $5 VISA gift card if your friend (maximum of 4 friends) completes this form and lists your email in the "Referred by" field. Your friends also get a $5 VISA gift card for signing up.
- If more than 4 of your friends sign up, we can only provide up to a $20 VISA gift card, but everyone you refer will still get a $5 VISA gift card just for signing up.
- It's important everyone signs up by the end of the day of Thursday, November 21. Once you get your gift card, be sure to use it right away!

Thanks for being a member of the CUTGroup! As usual, call or write with questions.

We signed up 100 new users through this campaign.

We now purchase gift cards with long lives (8 years!), though the value decreases from the card at 13 months after purchase, so you still have to carefully manage your gift cards.

Recruiting developers

One of the central tenets of the CUTGroup is that developers are a part of the process from start to finish. We also work with project managers of tech projects who want this feedback, but encourage them to still bring the developer to the test. That's why we never start planning a CUTGroup test unless someone with the product asks us to do so. This ensures that they are willing to learn from talking directly with users, and that they will make changes to the system when the results are in.

We engage developers on an ongoing basis by organizing, attending, and presenting at civic hack nights, and being an engaged part of Chicago's development community. Nothing can replace genuine engagement in your local development community, and a CUTGroup can be an important part (or start) of that engagement.

Design

Once we have a developer who wants to work with us, we set up a meeting to learn more about the product, decide how to segment the CUTGroup tester base so that we put together the best group for test night, and design the test itself.

Questions we ask at this stage:

1. What day do we want to do the test, and do we have a general part of town where we want to do it? This is based on availability of the developers and Smart Chicago staff. When it comes to location, we check to see where the most recent tests were conducted, and we endeavor to target a part of town where we haven't been in a while. There are some instances where an app is more applicable to a particular area as well. (We once did a test on Build It! Bronzeville which focused on the retail environment of a particular neighborhood, for example.)

2. What is the current stage of the app? Is it a paper prototype, is it in beta, or does it have thousands of users? This basic information gives us a start that helps the entire process. We've worked with apps in each of these states of being.

3. Who do we want to target? Do they have very specific requirements for the test? For instance, for Go to School!, we needed to find parents whose children attended a Chicago Public School and who took personal responsibility for getting those children to school in the morning.

4. What screening questions should we ask? For each test, we fashion some specific questions that help us screen testers to make sure we have a room of relevant people on test night. Examples are, "Do you write reviews on Yelp?" for a restaurant inspection site or, "Do you know your alderman's first name?" for a neighborhood news site. People who post on Twitter might be more likely to post on other forums. When testers sign up for CUTGroup, we do not capture much demographic information.

We want to keep it simple for testers to join the CUTGroup. Our screening questions are a way for us to capture more info that is relevant to the test.

5. What type of test? We employ a number of test types to get the information developers need: in-person test, remote test, and focus group are the main types.

6. What preliminary questions should we ask users? We also try to design questions for the test itself, starting with some framing questions. Sometimes it's an open question ("How do you use maps?"). For others, it's more specific ("Do you attend any of these types of meetings?"—with a list of six types).

7. Are there any questions specific to the app itself? This is where we get to the heart of the user interface questions that the developer wants answered.

8. What other goals does the developer have? This is where we cover the full circle of digital skills for the developer—what else do they want to get out of this test?

9. How and when will developers implement the feedback from CUTGroup testing? This is when we gauge how willing developers are to make changes and how we can help provide analysis and results to support them in making changes.

Here's a portion of an example test plan, done for Foodborne Chicago.

Example plan, Foodborne

Who are current users of Foodborne?
- http://blog.corynissen.com/2013/11/mapping-foodborne-chicago-reports.html

Who do we want to target for this CUTGroup test?

- We will be targeting Twitter users (with at least 100 tweets)—both very heavy and not as heavy users
- We are looking to get a mixed group from all areas of Chicago—we want to have approximately 2/3 of the group from the South Side, 1/3 of the group from the North and West sides
- We will ask for Twitter handles (not a required field)
- Types of screening questions to ask to gather more information:
  - Do you have a Twitter account? (required)
  - How often do you use Twitter?
  - How often do you go out to eat in a week?
- When creating our group, we will consider neighborhoods, Twitter usage, if they eat out, and if they have ever used 311 (mix of yes/no responses)

What type of test?

- Three sessions in a focus group style with small groups (5 people). We will be able to bring up pages and views to show the groups and get their feedback
- This is a qualitative test, since there will be a quantitative survey coming out later
- We will record the sessions

What do we want to ask? (Please note that these are just a rough outline of the type of questions we'll ask)

- How do you currently use Twitter? Do you tweet often? What types of things do you tweet about?
- If you got food poisoning, would you tweet about it? Would you tell people you're sick? Why or why not?

- If you get food poisoning, what do you do? Do you tell someone about it? Do you submit a 311 request?
- Show: Foodborne Twitter account and gather feedback about the tweeter and the tweet
- Show: Foodborne Chicago form page and gather general impression—What do you think this form does? How do you feel about this information?
- Discuss form information: Is this something you would fill out? Why or why not? Do you feel comfortable with giving this information? Do you trust this site?
- Do you know that this information is being sent to the Chicago Department of Public Health (CDPH)? How do you feel about that?
- Show: Q&A page—see how people feel about the request being made into a 311 request and the service tracker—this information might need to be more prominent elsewhere, this is where we will get more information about 311
- What would make you more willing to respond to a random tweet? Followup: Would a tweet from a person in the community be better?
- What would make you more willing to fill out this form?
- Do you like Foodborne Chicago? Why or why not?
- Do you think Foodborne Chicago is an effective way to report food poisoning? Why or why not?

What does Foodborne want to learn from the CUTGroup test?

- Foodborne wants to learn why South Siders are not responding
- How do residents feel about responding to a random tweet?

- Is the form too much? What fields might not need to be required?
- Would users feel better/worse if there were more CDPH or 311 visibility? (In terms of Twitter handles, domain names, and logos)

Segmenting

Once we've got a rough plan for the test, it's time to gather a relevant group of testers. We do this through MailChimp email campaigns integrated with Wufoo surveys in a three-step process:

- A **broad call-out** to people, giving them a general idea of the test type, asking testers their availability for a two-hour time period on the test date, and noting any specific requirements of the test
- An email discussing the **specifics of the test**, including location and test type, and asking them to affirm their availability, along with a preferred 30-minute slot in the context of the entire test period
- Emails to the set of people who **did not get segmented** into the test

The final steps in segmenting are done through a regular email service (not through the MailChimp mass email tool). We hand-schedule the testers into slots based on the tester's choice, the number of testers, answers to their screening questions, and test type. This is a great example of a task we'd like to automate with our Patterns software—we just haven't gotten around to it.

Here are some sample emails and surveys you can use in planning your own emails to testers:

Broad call-out, initial email

Subject line: Make $20 at an in-person test of a food poisoning app.

Hi <<First Name>>

We've got a new opportunity for you to make money in the Civic User Testing Group (CUTGroup) by testing an app.

This app helps report food poisoning incidents to the Chicago Department of Public Health. We want to know if this app serves your needs, and how it can be improved.

Are you available for a 30-minute, in-person test on Monday, April 14, between 4:00 and 7:30 p.m.? If so, you qualify. *Complete this form to start the process.*

We are in the process of gathering responses, so we will be in touch to check availability and let you know if you have been chosen. We are looking for 15-20 testers for this test.

For your participation, you will receive a $20 VISA gift card. You'll also help make better software for Chicago.

Thanks for being a member of the CUTGroup! As usual, call or write with questions.

Remember: if you want to be a part of this test, please complete this form.

Broad call-out screening questions

CUTGroup Test: Map Editor Website

If you would like to participate in the next CUTGroup test, please complete this form.

Email *

Are you available on Wednesday, January 22 from 4:30 – 7:30 PM? *
[Select one]

Have you ever edited a webpage? *
[Select one]

Have you ever made a comment on a Web site? *
[Select one]

What type of maps do you use? *

Do you use maps to explore your neighborhood? *
[Select one]

Do you use Foursquare.com? *
[Select one]

[Submit]

Complete text of this email here: https://smartchicago2012.wufoo.com/forms/cutgroup-test-map-editor-website/

Specifics of the test email

Hi <<First Name>>,

Thanks for your interest in our CUTGroup test of a food poisoning app. Now that we have our location nailed down, we'd like to know if you can still make it next Monday and, if so, what time you want.

Here are the details:

Methods

Monday, April 14, 2014
Between 4 and 7:30 p.m.
Blackstone Branch of the Chicago Public Library
4904 S. Lake Park Avenue

If you can still make it on this night, please let us know what time slot you want by *completing this form*. For your participation, you will receive a $20 VISA gift card.

There are limited slots available for this night, and it's coming up fast, so please let us know as soon as you can. We will get back to you with a confirmation if you are chosen to do this test.

If this location doesn't work for you, or if you just can't make it on this night after all, no big deal. If you have any questions or comments, just hit "reply" and let me know what you think.

Remember: use this form to tell us about your availability.

Thanks for being a member of the CUTGroup!

Specifics of the test survey questions

CUTGroup Test: Food Poisoning App

If you would like to participate in the next CUTGroup test, please complete this form. We will follow-up to check your availability.

Email *

Are you available for a 30-minute test on Monday, April 14 between 4:00 – 7:30 PM? *
[Select one]

Do you have a Twitter account? *
[Select one]

How many total tweets do you have on Twitter? *

Only if you want to, share your Twitter handle:

How often do you go out to eat in a week? *
[Select one]

[Submit]

Complete text of this email is here: https://smartchicago2012.
wufoo.com/forms/cutgroup-test-food-poisoning-app/

Did not get segmented email

Hi <<First Name>>,

I wanted to follow up with you about testing a food poisoning app *Foodborne Chicago*. We really appreciate your response to our call out.

Foodborne Chicago searches Twitter for tweets about food poisoning and then responds to those tweets with a link to report it so that the Chicago Department of Public Health can take any necessary action. Therefore, for this test, we were looking for heavy users of Twitter.

Since you indicated that you don't have a Twitter account, this just was not the right test for you.

We will send you info on our next test (should be in May or June), and we'll definitely try to spread the tests around so that everyone gets what they want out of the CUTGroup experience. We do keep track of these things!

Thanks for being a member of the CUTGroup. As usual, call or write with questions.

> Thanks for being willing to help out.
>
> Hi <<First Name>>,
>
> I wanted to follow up with you about testing a food poisoning app Foodborne Chicago. We really appreciate your response to our call out.
>
> Foodborne Chicago searches Twitter for tweets about food poisoning and then responds to those tweets with a link to report it so that the Chicago Department of Public Health can take any necessary action. Therefore, for this test, we were looking for heavy users of Twitter.
>
> Since you indicated that you don't have a Twitter account, this just was not the right test for you.
>
> We will send you info on our next test (should be in May or June), and we'll definitely try to spread the tests around so that everyone gets what they want out of the CUTGroup experience. We do keep track of these things!
>
> Thanks for being a member of the CUTGroup. As usual, call or write with questions.

The more we ask of testers, the smarter we get about testing

Overall, our strategy is to engage testers over multiple emails and gather new information. We develop deeper relationships and have more information that allows us to segment for new tests.

Our open rate for the blast availability emails, which are sent to all active CUTGroup users, regularly comes in at 45-55%, and we have a very low unsubscribe rate. People in the CUTGroup look forward to getting emails from us.

Now we have testers, a developer, and goals for our tests. Where should we hold the test? What do we do once we get there?

Scouting

An absolutely essential criteria is that CUTGroup tests take place all over. To decide where to hold our next test, we start with the list of Connect Chicago locations and other public computer centers in the county. Chicago Public Libraries have great, accessible, and free meeting room space for community organizations.

Much of the test location planning is classic event planning—it's important to make sure that librarians, patrons, security guards, and anyone else we encounter during the test is fully informed about and comfortable with our presence.

Proctoring

We use a mix of test administration styles in the CUTGroup: direct test proctoring, focus group, remote, and self-driven/in-person.

Direct test proctoring

This is the most common test style we use in the CUTGroup. Most often, we pair every tester with a single proctor who works with the tester to complete the test. Sometimes, based on the vagaries of scheduling, a single proctor will work with two or more testers at the same time.

Focus Group

Sometimes we can gather more intelligence about an issue in a focus group, which is great for getting qualitative information about a topic. We first have testers answer a survey, so that they have a chance to form their own opinions before the group discussion. Next, we engage them as a group—they all share their answers and we ask additional questions.

Remote

We always value in-person tests and believe they are an opportunity to convene participants from all areas of Chicago. We gain valuable responses in person and are able to record a tester's actions and reactions. But remote testing can give us the sheer quantity of answers that can put an exclamation point on test results.

Self-driven/in-person

These tests were created so that no proctors were needed. We designed this type of test before CUTGroup #4, EatSafe.co. It is an example of a test that had lots of testers, and some testers did not have a proctor with them at the time.

Analysis

After the test is done, we analyze the results and publish all of the raw test data as well. We export all of the results out of Wufoo into a spreadsheet, then use a Microsoft Word macro to populate an easy-to-scan document of the raw results.

The analysis is done by hand by the main proctor of the test. Sonja Marziano has done the majority of analyses for our CUTGroup tests, and she designed the method for exporting and formatting all test results.

Since the person writing the analysis was also present at the test itself, she is able to take the quantitative information and add qualitative insights so as to get actionable recommendations for the developers.

Followup

The interface recommendations are published in the body of our analysis blog post and we almost always create GitHub issues as well. This ensures that we create concrete, direct recommendations— nothing squishy or weak.

What happens next is up to the developer! Remember, a lot more can be found on our website and blog at smartchicagocollaborative.org. We've posted in-depth info and documentation on what we've learned.

> **DO YOU LIKE MONEY?**
>
> **Be a tester, get paid**
> The Civic User Testing Group (CUT Group) is a set of regular Chicago residents who get paid to test out civic apps.
> - Fill out a CUT Group profile and sign up to be a tester of civic apps, and we'll send you a $5 VISA gift card
> - If and when you are chosen to test a civic app, you get paid a $20 VISA gift card and bus fare.
>
> There is a large and growing community of "civic hackers" in Chicago technology developers who make websites, mobile apps, and other tools that often have specific use in Chicago. The goal is to make software that helps make lives better in the city. The problem is that lots of civic apps get attention among a smallish group of other developers and people interested in the world of open data, but do not get wide acceptance by the people they were made for regular residents of the city of Chicago.
>
> **You are going to change all that!**
> We need people from all over the city, using all sorts of devices, browsers and operating systems.
>
> **Smart Chicago | Civic User Testing Group**
>
> To receive your FREE $5 VISA gift card, simply sign up at: http://cutgroup.smartchicagoapps.org/

"The Money Flyer" was created by Emily Escarra and art directed by Kyla Williams. It drives home a key point about the CUTGroup.

References

Nightlife Magazine. (2013, February 19). Retrieved from http://mad.ly/571883

Amazon gift cards. (n.d.). Retrieved from http://www.amazon.com/gift-cards/b/ref=topnav_giftcert?ie=UTF8&node=2238192011&tag=donations09-20

ScripSmart. (n.d.). Retrieved from http://www.scripsmart.com/

Smart Chicago Collaborative. (n.d.). Gift Cards for CUTGroup. Retrieved from http://www.smartchicagocollaborative.org/gift-cards-for-cutgroup/

CUTGroup #9. (n.d.). Foodborne script. Retrieved from https://docs.google.com/document/d/1-qESfmC09h04H92pJeAM2JgCnTB-VkADTQaQcew1MXiU/edit

Kyla Williams prepares to proctor a test.

Smart Chicago Collaborative. (n.d.). Refer a friend. Retrieved from http://us5.campaign-archive2.com/?u=085247dea37361c-266002462c&id=d6771c1bc2&e=[UNIQID

CUTGroup #7. (n.d.). Everyblock iPhone App. Retrieved from https://docs.google.com/document/d/13RVFkWCVrA8CWXnYB3Xxz7py5Iqm-crc-4YtNMYyy8MA/edit?pli=1

Foodborne Chicago. (n.d.). Retrieved from https://www.foodbornechicago.org/

1871. (n.d.). Chicago's entrepreneurial hub for digital startups. Retrieved from http://www.1871.com/

Stats and things. (n.d.). Mapping Foodborne Chicago Reports. Retrieved from http://blog.corynissen.com/2013/11/mapping-foodborne-chicago-reports.html

CUTGroup. (n.d.). *New CUTGroup Opportunity: Test a Food Poisoning App.* Retrieved from http://us5.campaign-archive1.com/?u=085247dea37361c266002462c&id=d16b56c395&e=[UNIQID

CUTGroup test. (n.d.). *Map Editor Website.* Retrieved from https://smartchicago2012.wufoo.com/forms/cutgroup-test-map-editor-website/

CUTGroup test. (n.d.). *Update on the next CUTGroup Test of a Food Poisoning App.* Retrieved from http://us5.campaign-archive1.com/?u=085247dea37361c266002462c&id=cod11ccfc9&e=[UNIQID]

Connect Chicago. (n.d.). Retrieved from http://weconnectchicago.org/

Chicago Public Library. (n.d.). *Book a Meeting Room.* Retrieved from http://www.chipublib.org/book-a-meeting-room

Examples

Based on how we've put together this book, we may have given you the idea that we've put our methods into practice in an utterly linear and structured way. This would be giving you the wrong impression.

In reality, the CUTGroup is just one of many programs we have going on at any given time. As you can see from the Acknowledgements, we've worked with dozens of experts giving us bits of advice and helping us figure out what we're doing.

The best way we can show you how we got to where we are is to directly show you how our methods met practice. We cover some of the tests we've done below and talk briefly about the impact each had on our methods.

For each of the examples, we have a complete set of results published on the Smart Chicago website. This includes links to every piece of communication we've sent to testers, details on all test objectives, and the raw results of every test.

#1 — FreedomPop Router

The first test we ever did was also the one that was least like the others. We spent two months recruiting testers, and we were ready for our first test. By this time, the City of Chicago was looking to launch a pilot with FreedomPop, but they weren't sure if the 4G routers they offered would be of acceptable quality. Since expanding access to the Internet is one of the core missions of Smart Chicago, we decided to come up with a process to use the CUTGroup to test the hardware, customer workflow, and coverage for this product.

Segmenting was important in this test, because the product was only available at this price in certain ZIP codes of the city. At the end of March and beginning of April 2013, two emails were sent to CUTGroup members in the twenty ZIP codes where the lowest-cost EveryoneOn option was available.

We had 94 possible testers, and 8 people signed up for the test through what would soon become our standard process for recruiting and segmenting. This was one of two tests that were done remotely. While we learn a lot from remote testing, we gain even more insight from talking with testers in person.

Testers indicated that unboxing the device was a breeze and that it worked well. We were able to determine that the service met expectations by engaging with people over email, home delivery, and a web form. The system worked, but we knew we had to get people together in a room to start delivering on the promise of community engagement.

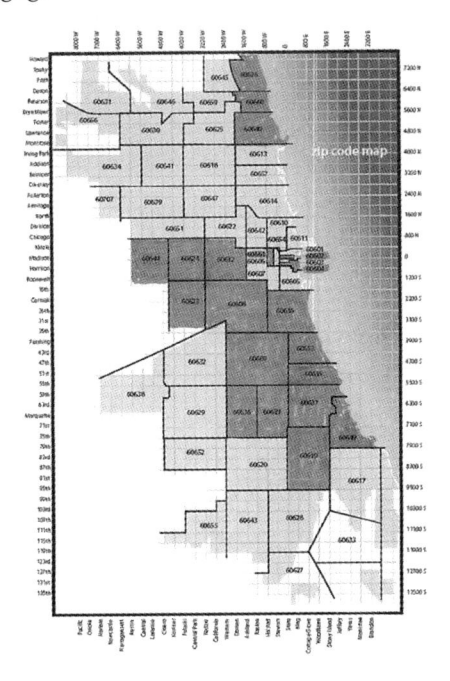

We had to draw testers from just 20 Chicago ZIP codes.

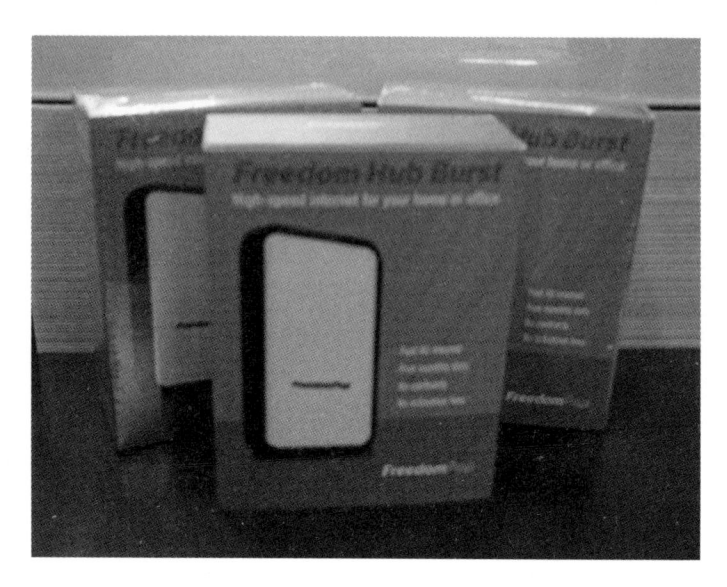

The Freedom Pop 4G Routers

References

Escarra, Emily. (2013, July 26). Results of CUTGroup001: FreedomPop Router Test. Retrieved from http://www.smartchicagocollaborative.org/results-of-cutgroup001-freedompop-router-test/

Mayor's Press Office. (2013, April 16). Mayor Emanuel Announces New Campaign to Expand Affordable Internet Options for More than 1 Million Chicagoans, Retrieved from http://www.cityofchicago.org/city/en/depts/mayor/press_room/press_releases/2013/april_2013/mayor_emanuel_announcesnewcampaigntoexpandaffordableinternetopti.html

#2 – Go to School!

This test centered around a website that promised "Four steps get to your CPS school on time." It is a simple wayfinding tool with school start times and contact information baked in. We learned two things from our first in-person test: how to segment users for a complex requirement and the power of the "shared screen" in community engagement.

We had to do a significant amount of segmenting of our list of 368 CUTGroup members. We asked all of them two questions:

- Are you responsible for taking one or more children to a Chicago public school this fall?
- Are you available for testing on either May 28 or May 30?

There were 28 people who responded to these requirements. Based on the locations of these people, we decided to hold tests at two libraries: the Kelly Library in the Englewood neighborhood on Tuesday, May 28th, and at the Uptown Library in the Chicago Uptown neighborhood on Thursday, May 30th. There were 14 people who expressed interest in each location, so it worked really well.

The testers were spot-on knowledgable about the topic of getting kids to school on time. In an effort to maintain privacy and to provide the widest possible utility of our testing to civic developers everywhere, we developed some shorthand "personas" that helped in identifying their stance vis-a-vis the app.

This was the first time I personally came into contact with the idea that the CUTGroup was a great framework for discussing difficult topics in unemotional and (hopefully, ultimately) helpful ways. I proctored "Tester #2: Father responsible for three children (17, 16, and 10) Samsung Galaxy Kelly Library, Englewood." The topic of school closings and "safe passage" was very active and contentious. But in the context of reviewing this site, he was able to express in everyday terms the choices he made in terms of danger associated with certain routes.

This is the joy and the promise of the CUTGroup. We end up using a "shared screen"—a concrete, tangible interface that everyone can point to, and talk about—to serve as a common ground for discussion of difficult topics that can affect policy.

On a more prosaic note, we were able to uncover trouble with the custom time/date picker screen. People didn't understand how to work it. The developer, Tom Kompare, made changes to the picker based on the feedback.

This test was an early example of the kind of broad geographical range we can deliver in our tests.

Julie Harpring proctors a test while Chris Gansen takes notes.

References

O'Neil, Daniel X. (2013, August 13). *Results of CUTGroup002: Go-2School User Application Test.* Retrieved from http://www.smartchicago-collaborative.org/results-of-cutgroup002-go2school-user-application-test/

Connect Chicago. (n.d.). *Kelly Library.* Retrieved from http://locations.weconnectchicago.org/location/kelly-library-6151-s-normal-boulevard

Connect Chicago. (n.d.). *Uptown Library.* Retrieved from http://locations.weconnectchicago.org/location/uptown-library-929-w-buena-avenue

O'Neil, Daniel X. (2013, July 21). *CUTGroup #2, Tester #2 Father responsible for three children (17, 16, and 10) Samsung Galaxy Kelly Library, Englewood.* Retrieved from http://vimeo.com/70725523

Go To School! (n.d.) Retrieved from http://cps.go2school.org/

O'Neil, Daniel X. (2013, August 12). *Go2School Screen 2/4 -- Date and Time Picker.* Retrieved from https://www.flickr.com/photos/juggernaut-co/9495316852/

Kompare, Tom. (n.d.) *@tomkompare.* Retrieved from https://twitter.com/tomkompare

O'Neil, Daniel X. (2013, July 7). *News story: Finding 'real people' to test civic apps.* Retrieved from http://www.danielxoneil.com/2013/07/07/news-story-finding-real-people-to-test-civic-apps/

#3 – Chicago Health Atlas

The Chicago Health Atlas is a place where you can view citywide information about health trends and take action near you to improve your own health. The test took place at the Erie Family Health Center, a Smart Health Center in Humboldt Park.

We wanted to see how everyday residents were using it and make sure that people could find what they were looking for on the site.

The test revealed a number of user interface issues, mostly revolving around a key feature: the ability to see healthcare resources near the user. We made some simple modifications to the site, mainly making the text bigger, explaining the features more clearly, and showing links to resources as users were viewing the map. We've yet to re-test the site, but informal testing (watching people view the site) indicates that we've improved clarity.

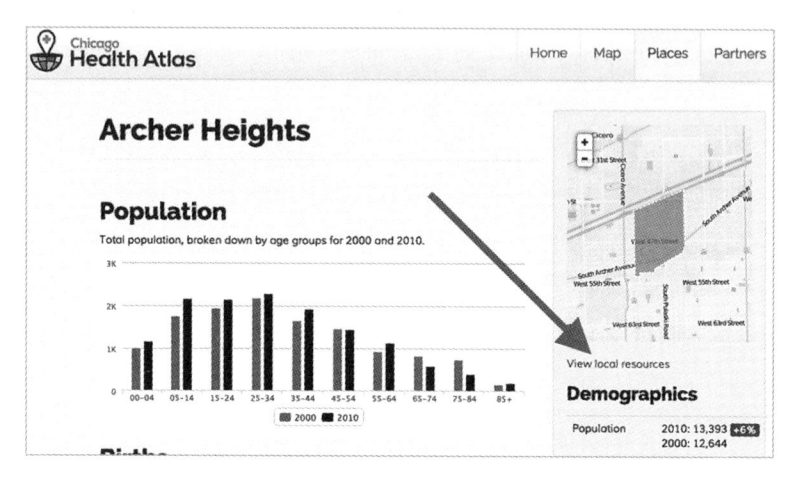

The "View local resources" link was missed by many users.

A tester reviews a detail page.

References

Whitaker, Christopher. (2013, August 6). CUTGroup Test #3: Chicago Health Atlas. Retrieved from http://www.smartchicagocollaborative.org/cutgroup-test-3-chicago-health-atlas/.

Smart Chicago. (n.d.). Smart Health Centers. Retrieved from http://www.smartchicagocollaborative.org/projects/smart-health-centers/

Chicago Health Atlas. (n.d.) Retrieved from http://chicagohealthatlas.org

#4 – Eatsafe.co

Our fourth session was focused on EatSafe.co, a website that shows details of food inspections near you, which was developed by Hoy Publications. This in-person test took place at the Chicago Public Library's Hall Branch at 4801 S. Michigan Ave in the Grand Boulevard neighborhood.

This was one of the more community-focused tests. We had overbooked the sessions, and lots of people showed up early. It was touchy for a while because we were struggling to match up proctors to testers. After we loosened up and accepted the fact that we had to double- and triple-up testers to proctors, people really enjoyed working together and talking about the website. Fernando Diaz, Hoy Managing Editor, had this to say:

> "Partnering with the CUTGroup was the most effective research experience I've ever been a part of. We met real Chicagoans who were generous with their time and feedback. And among the highlights is that we have all of the results for further evaluation and incorporation into future iterations of our project."

We got good, actionable feedback from testers. When clicking a violation, testers wanted an explanation of the violation. (The current function was that they were directed to a list of establishments with the same violation.) Testers were also interested in better explanations of the inspection results.

Examples

We had one particularly remarkable experience worth relaying, because it shows how civic apps fit into the lives of residents. The tester was the mother of a child in a public school near her home. While browsing restaurants, she saw that her child's school had failed an inspection.

She said: "I didn't hear anything about it. I didn't know." You could hear the frustration. We've all been in situations where a lack of information takes us aback, makes us feel powerless. She read the text of the violation, relating to washing facilities. "That's not good. That's not a good look at all," she said.

I asked her what she wanted to do next, asking a classic UX testing nudge, drawing the tester back to the interface. She wanted to "contact the school, and find out what's going on, because my son is eating lunch there." There was no way to contact the school from the website.

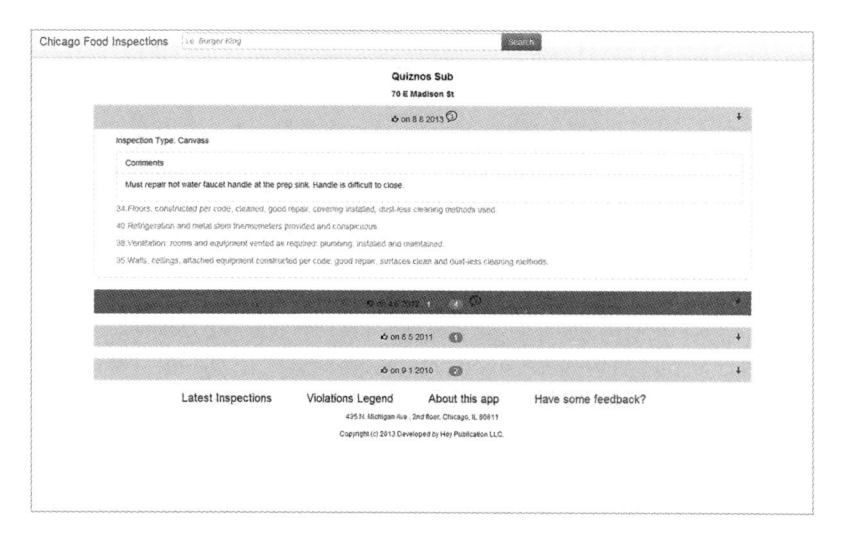

Detail page for inspection result. Testers had difficulties searching for a specific establishment and deciphering the results.

Daniel X. O'Neil observes a tester with piano and candy in background.

References

Marziano, Sonja. (2013, November 21). CUTGroup Test #4: EatSafe.co. Retrieved from http://www.smartchicagocollaborative.org/cutgroup-test-4-eatsafe-co/

Hoy Publications LLC. (n.d.). Chicago Food Inspections. Retrieved from http://www.eatsafe.co/

City of Chicago. (2011). Food Inspections. Retrieved from https://data.cityofchicago.org/Health-Human-Services/Food-Inspections/4ijn-s7e5

Vivelohoy. (2014). Retrieved from http://www.vivelohoy.com/

Smart Chicago. (2013, December 26). CUTGroup 4 Eastsafe Tester 16. Retrieved from https://www.youtube.com/watch?v=Sk_j7xppUtg#t=14

Smart Chicago. (n.d.). Make $20 at an in-person test of a food inspection website. Retrieved from http://us5.campaign-archive2.com/?u=085247dea37361c266002462c&id=63604c0cf7&e=[UNIQID

CUTGroup Test: Food Inspection Website. (n.d.). Retrieved from https://smartchicago2012.wufoo.com/forms/cutgroup-test-food-inspection-website/

#5 – ChicagoWorksforYou.com

Our first remote test was for ChicagoWorksforYou.com, a Smart Chicago project. ChicagoWorksforYou provides citywide information about city service requests. Users are able to get a ward-by-ward view of service delivery in Chicago, learn about the top service requests made on a given day, view photos of requests, and learn more about the process of submitting service requests.

We did this test remotely because we wanted to get as many people as possible, mainly because we were in a bind—we had to dispense gift cards that were about to expire, and the Thanksgiving holiday was coming up. But we also had a theory that the CUTGroup could serve as a form of advertising and marketing for a website and thereby increase the user base. (As time has passed this has not proved true.)

We started by sending out an email to all 565 CUTGroup participants, asking them if they would be available to provide feedback through a remote test. We received 116 responses in one day and asked 90 random respondents to do the test.

We asked more "yes" or "no" questions than we usually do, in order to gather quantitative results. We leavened this with open-ended questions to see what users were interested in when visiting the site, and finally we asked users to click on specific links of the website, and discuss their experiences. We were pleasantly surprised at the thoroughness of testers' responses without a proctor being present.

In addition, we asked 5 willing testers to do a test via screen share. We randomly chose our respondents, compiling a group of testers from all areas of Chicago, and had a variety of responses to our questions. Due to technical problems, we were unable to do the screen share. We should have tested it more and provided more technical instructions on how to screen share. The other issue was more conceptual: can we do the same kind of community engagement in a remote test, over a shared video connection, that we can do in a public computer center? We're going to look into this more.

We promise in our release form that, "In any report we might publish, we will not include any information that will identify you." So in order to keep track of testers, we ask them to provide a "tester profile," which we describe as "a short phrase that describes yourself and/or your relationship to the software." People have fun with the tester profiles.

Here are a couple of responses we heard from our testers specifically about the remote test:

> "An online test is a better form of testing a website or app. It is done within the comfort of one's home, with flexible times. It allows more people to participate and allows for a more natural environment." – Tester #77, Elizabeth07

> "I do like the remote survey better than the one I had to go to the library for. The particular public library I had to go to was in a very dangerous area and I didn't know before I went... Other than that it has been a pretty good experience being a part of the CUTGroup thus far and I'm definitely willing to give my input on multiple websites." Tester #46, 3rd Year Student

Here's the complete text of our release form.

CUTGroup Consent To Participate In Usability Test

What is the purpose of this test?

We are asking you to participate in a usability test because we are trying to learn more about how people are likely to use the website you are being asked to test.

How much time will this take?
This study will take about 30 minutes of your time as follows: 5 minutes of discussion, 25 minutes for evaluating the site and 5 minutes of wrap-up.

What will I be asked to do if I agree to participate in this study?
If you agree to be in this study, you will be asked to perform basic interactions with the application under consideration.

The test session will be video recorded. Following the session, the video recordings may be analyzed by the CUTGroup in order to provide further feedback, and videos with voice and the website interface may be shared to educate others about website usability.

What are the risks involved in participating in this study?
Being in this study does not involve any risks other than what you would encounter in daily life interacting with a computer-based application. It is important that you understand that your information will remain confidential during and after the testing session.

What are the benefits of my participation in this study?
You will not personally benefit from being in this study beyond the basic remuneration that has been offered. However, we hope that what we learn will help contribute to improving the quality of the applications being studied.

Can I decide not to participate? If so, are there other options?
Yes, you can choose not to participate. Even if you agree to be in the study now, you can change your mind later and leave the study. There will be no negative consequences if you decide not to participate or change your mind later.

How will the confidentiality of the research records be protected?

The records of this study will be kept confidential. In any report we might publish, we will not include any information that will identify you. Study records will be stored securely and only the CUTGroup will have access to the records that identify you by name. Some people might review our records in order to make sure we are doing what we are supposed to. If they look at our records, they will keep your information confidential. Digital versions of all video recordings will be kept in password-protected files and will be destroyed within three years after start of the study.

Whom can I contact for more information?

CUTGroup: 312.565.2867

You will be given a copy of this information to keep for your records.

Statement of Consent:

I have read the above information. I have had all my questions answered. (Check one:)

☐ I consent to be in this study.   ☐ I DO NOT consent to be in this study.

Signature:_____

Date: _____

Printed name: _____

Personal Release for Filming:

I authorize the CUTGroup to take and use video recordings of me in connection with the usability study.

Signature: _____

Date: _____

Printed name: _____

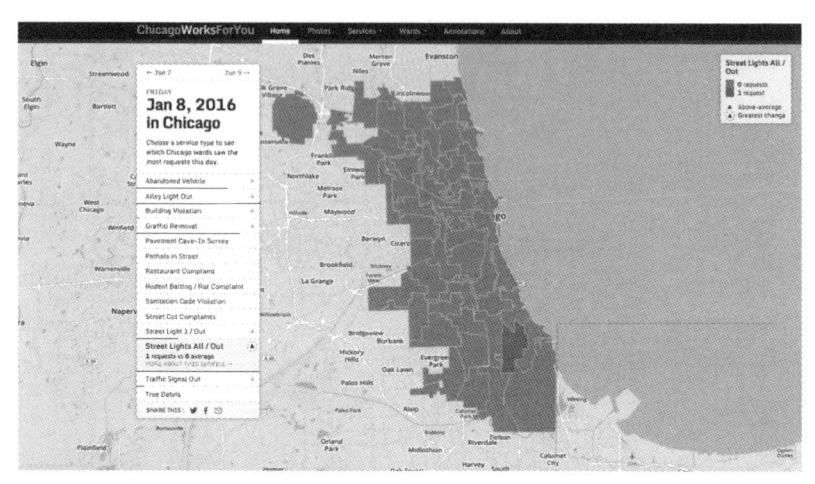

The Chicago Works For You homepage.

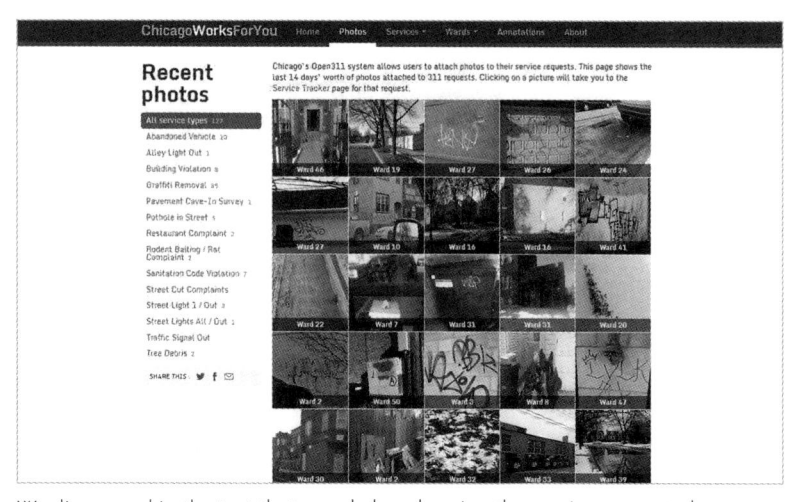

We discovered in the test that people loved seeing the service request photos, so we decided to bump that up in the interface and expand the section.

References

Marziano, Sonja. (2013, November 25). CUTGroup #5: ChicagoWorks-ForYou.com. Retrieved from http://www.smartchicagocollaborative.org/cutgroup-5-chicagoworksforyou-com/

Chicago Works for You. (n.d.). Retrieved from http://chicagoworksforyou.com/.

City of Chicago. (n.d.). Open311 API Documentation. Retrieved from http://dev.cityofchicago.org/docs/api

Smart Chicago. (n.d.). Make $20 at a remote test of a city service delivery Web site. Retrieved from http://us5.campaign-archive1.com/?u=085247dea37361c266002462c&id=3937c108a4&e=[UNIQID]

The Chicago Boundary Service. (n.d.). Retrieved from http://boundaries.tribapps.com/

#6 – OpenStreetMap Editor

For our sixth test, we focused on the editor feature of OpenStreetMap.org. This in-person test took place at the Chicago Public Library's Rogers Park Branch at 6907 N. Clark Street in the Rogers Park neighborhood.

OpenStreetMap (OSM) is a website that is built by a community of mappers who contribute local knowledge and information to a map for everyone to use. Anyone can sign up, add information, and edit the map.

OSM is open data, and you are free to use it for any purpose as long as you credit OSM and its contributors. The main thing we tested was the ease of signing up and editing a map. This is of enormous interest to us at Smart Chicago, because we think that OSM is an opportunity for community members to describe their own streets, buildings, and assets with greater accuracy than anyone else. The fact that OSM runs millions of map experiences per day means that the more regular residents update OSM, the more communities can put their best foot forward on the web. The more comfortable people are in editing the map, the more accurately that communities can be shown on the map.

Sixteen testers provided their feedback regarding OSM, and we learned a lot. Not only about the functionality of the map editor, but also about people's feelings on the concept of editing a map. Some testers liked the idea of contributing knowledge for others to use, while other testers thought the concept of making live changes to a map was "scary" or "dangerous." Here are a couple of telling comments we got:

> *"I believe in power of people and having a significant contribution to these things. Gives a sense of community and add value in the sense of belonging"* – Tester #5, kirehernan

> *"I personally would but I would not want others to have the same access as I would due to the lack of restrictions."* – Tester #7, B

We also saw that the map scale posed challenges to the testers. Some became frustrated that when their search was not in the visible map area, they had to choose "Search Worldwide."

When searching, testers sometimes received too many options that were irrelevant to what they were looking for, or they could not find their search term. Testers are interested in having a clearer way of searching for locations.

Fourteen testers (88%) said they liked the site, and 11 testers (69%) said they would use the map editor again. Here are some reasons why testers wanted to use OSM's editor:

- Update outdated data
- Contribute to a neighborhood's visibility
- Feel the need to customize or make a map more personal
- Add information about safe pedestrian and bike paths

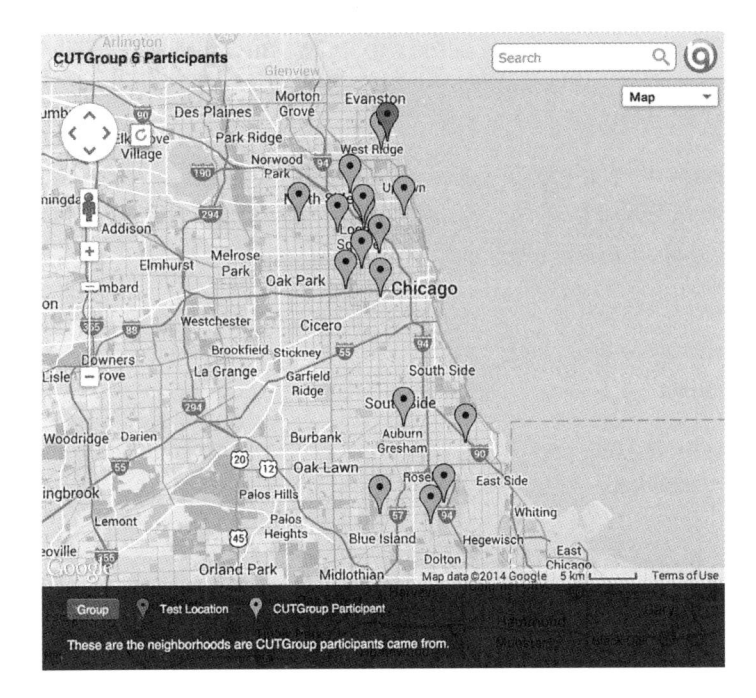

The geographic range of testers, for a test given in the harsh winter weather, really surprised us.

Ian Dees works with a tester at the Rogers Park Branch of the Chicago Public Library.

Examples

References

OpenStreetMap. (n.d.). Retrieved from http://www.openstreetmap.org/#-map=5/51.500/-0.100

Marziano, Sonja. (2014, January 22). CUTGroup #6: OpenStreetMap Editor. Retrieved from http://www.smartchicagocollaborative.org/cut-group-6-openstreetmap-editor/

Connect Chicago. (n.d.). Rogers Park Library. Retrieved from http://locations.weconnectchicago.org/location/rogers-park-library-6907-n-clark-street

Open Street Map. (n.d.). Copyright and License. Retrieved from http://www.openstreetmap.org/copyright

#7 – EveryBlock iPhone App

EveryBlock was coming back. The popular Chicago version of the neighborhood news and conversation website was relaunched in January 2014 by new owners, Comcast NBC Universal. They wanted to test their iPhone app—a vestige of the old site—to help them plan new features after their relaunch. I was part of the original team that launched EveryBlock in 2007 and had continued working in civic tech at Smart Chicago, so it was fun to get a chance to test a product I had worked on long ago but which still had relevance to me in civic tech.

This in-person test took place at the Chicago Public Library's Mayfair Branch at 4400 W. Lawrence Avenue in the Mayfair neighborhood.

We sent out an email to 269 CUTGroup participants who said they had an iPhone as their primary or secondary device. We asked them if they would be willing to test a neighborhood app on February 10, 2014. We also asked some screening questions to gather extra information, and we chose our group of participants based on a diverse selection of answers.

We had 12 testers who came from neighborhoods across Chicago including Albany Park, Hermosa, Edgewater, Uptown, Logan Square, Auburn Gresham, and more. The farthest a tester traveled from their home location was 15.3 miles.

This in-person test was the first opportunity we had for every tester to be paired with a proctor. In previous tests, some testers were paired with proctors, while others would answer questions about the website or app through an online form.

One thing we tested was the propensity of testers to want to post via the EveryBlock mobile experience, and 83% of testers said they would. The majority of testers thought this was a convenient option, and said they wanted to comment on things while they are happening. We noted that posting from mobile was a far more common activity than even a year ago, when the site was shut down.

The biggest takeaway from this test is that users were interested in features which would allow them to have an experience that matched their experience on the EveryBlock website.

A neighbor message used during the EveryBlock CUTGroup test.

Examples

EveryBlock Chicago returned in January 2014.

References

EveryBlock. (2013). Farewell, Neighbor. Retrieved from http://goodbye.everyblock.com/

Connect Chicago. (n.d.). Mayfair Library. Retrieved from http://locations.weconnectchicago.org/location/mayfair-library-4400-w-lawrence-avenue

#8 – Waitbot

For our eighth CUTGroup session, we tested the Waitbot app, where you can find estimates for waiting times of all sorts of things, including transit, restaurants, airports, and more. This test had an in-person and a remote component to it. The in-person test took place at the Chicago Public Library's Clearing Branch at 6423 W. 63rd Place in the Clearing neighborhood.

Through this test, we were interested in finding answers to these questions:

- What makes users download an app? Delete an app?
- Do users want to use Waitbot on a daily basis? Why or why not?
- What features do users want?
- What other wait-related categories would users want to see?
- Do users want to share wait-time information on social media?

On March 5, 2014, we sent out an email to all of our 749 CUT-Group participants. We asked them if they would be willing to test a wait-time estimates app on March 12. We asked some screening questions to gather information, and we chose our group of participants based on a diverse selection of answers and also device types.

We were interested in having about 15 participants from different Chicago neighborhoods, but we only had 6 testers come to the test in person. A lot of testers could not come due to a combination of weather and distance, so we reached out to 4 more testers to do the test remotely.

For the in-person test, proctors were able to work with testers one-on-one. Testers looked at the app on their own devices and provided feedback, while the proctors wrote down notes. After the test, we sent out additional, optional questions by email to see if testers were using the app and to see how they liked the app in their own neighborhood.

For the remote test, we asked testers to use the app on their own, and we provided questions to lead them through the test. In the end, we got great responses from both types of tests.

Most testers were not interested in sharing wait-time information on social media. One tester would share on Facebook only if it was automatically connected, while another tester said he would not do it unless there was an incentive. Only 3 out of 10 testers would share on social media.

When testing the Waitbot app, testers liked the transit page and the fact that it populated with nearby options. There was some confusion with color-coding, and testers wanted added features such as route display. However, testers overwhelmingly liked this page.

One tester, My eyes are dried out (#10), explains why he doesn't like the Waitbot app in general, but thought that the transit page was the most useful:

> "The Swiss army knife is useful and practical. Then the impostors 'improved on it,' making it bigger and more cluttered with useless features. Sometimes I feel app creators want to entice a large crowd, instead of just perfecting one good thing."

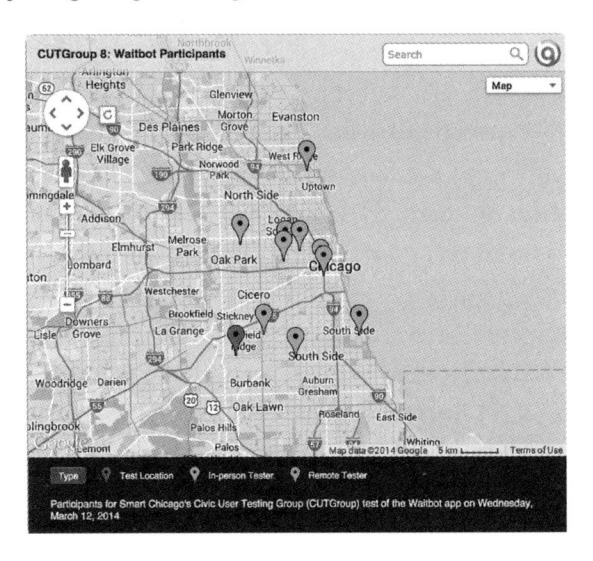

We had a mix of in-person and remote testers.

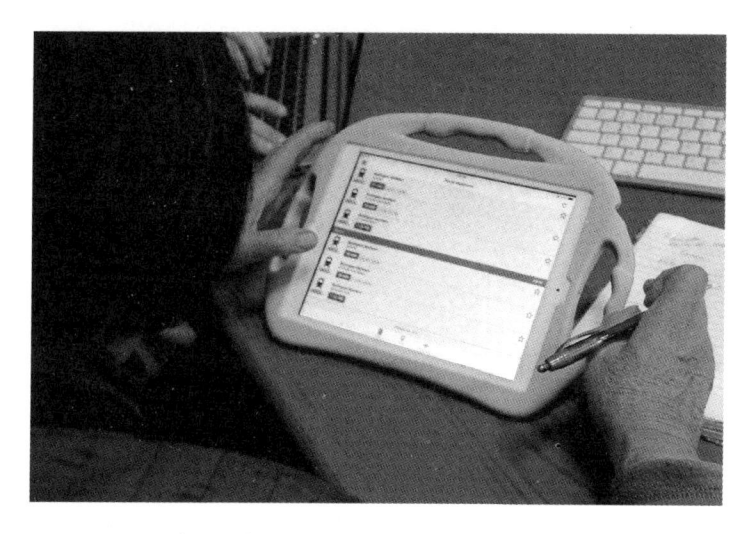

A tester reviews Waitbot on her own device.

References

Marziano, Sonja. (2014, March 12). CUTGroup #8: Waitbot. Retrieved from http://www.smartchicagocollaborative.org/cutgroup-8-waitbot/

Waitbot. (2013). Retrieved from http://waitbot.com

Connect Chicago. (n.d.). Clearing Library. Retrieved from http://locations.weconnectchicago.org/location/clearing-library-6423-w-63rd-place

#9 – Foodborne Chicago

For our next test, we tested Foodborne Chicago, an app that searches Twitter for tweets related to food poisoning and helps users report these incidents to the Chicago Department of Public Health. Joe Olson, one of the creators of Foodborne, received a grant, administered through Smart Chicago and the Chicago Community Trust, to build better communication strategies to engage all Chicago residents.

Cory Nissen, a statistician who wrote the Twitter classification code for this project, mapped out Foodborne Chicago reports and showed there were fewer responses from the South Side of Chicago.

That is to say, of all the people who publish a tweet that includes the phrase "food poisoning" from inside the city of Chicago, those on the South Side are less likely to click on Foodborne's prompt and complete a report. Through this test, we were interested in learning more about how people use Twitter, and whether there were differences among communities and networks.

This was also the test where we had the largest number of direct insights that could be turned into features and GitHub issues to improve the site. We made a number of changes, and the results are promising.

Here is a list of questions we wanted to answer through this test:
- How do users feel about responding to a random tweet?
- Does the form require too much information?
- Would users feel better or worse if there was more Chicago Department of Public Health or 311 visibility?
- Why might residents on the South Side of Chicago not respond as often as residents from other Chicago neighborhoods?

On April 9, we sent out an email to all of our 754 CUTGroup participants. We asked them if they would be willing to test an app on April 14, 2014, that helps report food poisoning incidents to the Chicago Department of Public Health via 311. We asked potential testers some screening questions and then chose a group of participants who were Twitter users, and who were primarily from South Side neighborhoods. This in-person test took place at the Chicago Public Library's Blackstone Library at 4904 S. Lake Park Avenue in the Kenwood neighborhood.

We did a focus group-style test with three 30-45 minute sessions, with 5 people in each session. First, we had testers fill out a survey, prepping them to form their own opinions before the group discussion. Next, we asked testers to discuss their answers, and then we asked additional questions as part of a group discussion. We had some very interesting in-depth conversations, not only about Foodborne Chicago, but also about Twitter and social media use in general.

We learned that Twitter is used as a private communications network for the majority of our testers, even though their tweets are public. Twitter users tend to connect with people they know or who are in their networks. Most testers were not sure about responding to people they did not know, because strangers tend to be spammers.

Foodborne prompts people on Twitter to fill out a web-based complaint form that routes to 311. For this user test, we wondered whether the 311 complaint form had too many fields. Would users want to fill it out? We learned that users had little issue with the form, and they liked that it was simple and did not require too much information. The tester Bakunin (#2) thought that it was his "civic duty" to fill out the form to make sure others did not get sick. While we were not asking too much information on the form, testers still had questions about our privacy policy and the process.

Overall, testers were interested in seeing a stronger connection to the City of Chicago, the Chicago Department of Public Health, and 311. Testers thought that seeing relevant logos on the website would emphasize these official, municipal connections. Testers also indicated that language about "The City" on the form was not descriptive, and they wanted more information about the process.

Based on feedback from the test, we immediately changed our tweet language to include compassion and an official component— the Chicago Department of Public Health. We also added a Twitter card (a way to attach rich photos, videos and media to tweets) in order to provide more detail about information being sent through the Chicago 311 service. Twitter cards weren't available to small publishers like us when we first launched Foodborne, so that is a lesson in itself— we should be on top of new features and implement them as soon as we can.

We changed the URL to *https://www.foodbornechicago.org/* from *https://foodborne.smartchicagoapps.org*. People said they would be more likely to click that link.

Testers felt more comfortable when they knew that there would be an official response. We changed the header information to include the Chicago Department of Public Health name and logo.

Testers also felt more comfortable with the form and the process of providing their information after reading the Q&A page. The final step is figuring out ways to be part of more people's networks so that they feel comfortable clicking the link in a Foodborne tweet and submitting the form to report on their food poisoning.

Twitter cards were effective in showing the quasi-official nature of Foodborne Chicago.

Foodborne Chicago has been the subject of lots of press attention, which helps in overall success. The more familiarity people have with the site, the more likely they are to complete the form.

References

Marziano, Sonja. (2014, June 18). CUTGroup #9 – Foodborne Chicago. Retrieved from http://www.smartchicagocollaborative.org/cutgroup-9-foodborne-chicago/

Knight Foundation. (n.d.). Chicago Food: Smart Chicago Collaborative. Retrieved from http://www.knightfoundation.org/grants/201347651/

City of Chicago. (n.d.). Department of Public Health. Retrieved from http://www.cityofchicago.org/city/en/depts/cdph.html

Nissen, Cory. (2013, November 25). Mapping Foodborne Chicago Reports. Retrieved from http://blog.corynissen.com/2013/11/mapping-foodborne-chicago-reports.html

Foodborne. (n.d.). Add Twitter Card support. Retrieved from https://github.com/smartchicago/foodborne/issues/49

Foodborne. (n.d.). Change header information to include Chicago Department of Public Health. https://github.com/smartchicago/foodborne/issues/54

Twitter Cards. (n.d.). Retrieved from https://dev.twitter.com/cards/overview

#10 – Build It! Bronzeville

Build It! Bronzeville is mobile game app that enables players to experiment with urban development, based on real-life places while using GIS technology. The developers won first place at the Center for Neighborhood Technology 2013 Urban Sustainability Apps Hackathon competition, and a CUTGroup test was part of the prize package. We were looking forward to working with a project that was in a very early, alpha version, something we'd never done before. It was also the first time we tested a game app.

Our in-person test took place at one of the Connect Chicago locations – Chicago Public Library Chicago Bee Branch in the Bronzeville neighborhood, which we chose because the game uses data on vacant lots in Bronzeville. Players might, for example, enhance safety, improve visual appeal, and increase foot traffic on virtual versions of real streets. Vacant properties in the game are based on real, vacant neighborhood land in Bronzeville.

We were excited to learn about how testers play games and what aspects they enjoyed the most. We were also very interested to learn whether testers appreciated the community-improvement element of this game.

With help from Smart Chicago, the development team proctored the test (a fact that we didn't reveal to testers, so that we'd be sure to get unbiased feedback). On test day, we had a fairly small group showed up. We confirmed 14 testers but only 6 came. Our testers tend to communicate with us often via email and phone to let us know when they are not coming, but this time, we were not sure why we had such a big no-show rate. The testers who did show up came from across the city, although we'd originally wanted to recruit heavily from Bronzeville because of the local focus.

Here's what we learned: testers definitely enjoyed the app's community improvement aspect, and they liked adding new, virtual buildings and businesses to the neighborhood. It felt like a local, custom version of Simcity. It was also useful for testers to see the flow of the game and to give feedback on the app's design elements. (Developers used simple graphics for Level 1 and more detailed graphics for Level 2, giving testers a sense of both possible aesthetics.)

One aspect of the app proved challenging, though. Players were asked to submit real receipts from purchases at local businesses, which the app converts into in-game benefits. Ultimately, receipts will be collected from the entire community, stripped of all personally identifying information, and developed into a report that community members can use to better understand how their local economy works.

But the technical aspect of the receipt submission feature wasn't ready at testing time. Testers had to use their phone's camera to take a photo of the receipt, and they were reluctant to share these without understanding exactly how receipts would be used. Testers said they'd prefer to "check in" at local businesses or organizations, such as libraries or community centers, instead.

Another issue: Because it was an alpha test, the app just didn't work on some Android phones. There were a lot of bugs, the keyboard screen crashed several times, testers were kicked out of the game, etc. Looking back, it might have been better to push back the date of the CUTGroup test until these basic bugs were fixed. Despite the challenges, we were still able to get enough good feedback to help Team Build it! with the direction of their app.

References

Team Build It! (2015) Retrieved from: http://teambuildit.net/

Tester Music Lover (#1) describes this game as "Edutainment–educating and entertaining at the same time."

#11 – Expunge.io

Expunge.io is a youth-led app that addresses a critical problem affecting tens of thousands of youth arrested each year: juvenile records. It grew organically out of a series of conversations between Mikva Challenge Juvenile Justice Council (JJC) youth, involved in an initiative called #CivicSummer in 2013, a summer program for youth to become involved in Chicago through civics, media and technology.

Although most expungement requests are granted, the process is complex. It means navigating a labyrinth of bureaucracy, interacting with intimidating institutions like the police department, and filing forms with the Clerk of the Court's office. In Cook County alone in 2012, there were 25,000 youth arrested but only 70 expungements.

Expunge.io, driven by the vision of the JJC, is an app that was created to change all that. It's a website designed for people with juvenile records in Illinois to kick off the process of expunging those records. More than that, we wanted Expunge.io to be helpful to everyone, regardless of whether or not they have a record. Information travels through personal networks. We wanted the word to spread.

Throughout Expunge.io's development, Smart Chicago acted as a connector, bringing partners, resources, and ideas into the ecosystem around this app. We also wanted to ensure that Expunge.io still had a youth voice after it was built, and Chris Rudd of Mikva Challenge asked us to do focus groups around that, as part of a grant he had received to support the project, from the Knight Foundation. We realized that the CUTGroup methodology would work really well here.

For these tests, we didn't do our normal call-out to testers, working exclusively with Mikva Challenge youth instead. (It worked well—many of our Expunge.io testers signed up to participate in other CUTGroup tests, too.) Testers readily helped us identify some things to change. For example, they wanted more upfront information about costs. Even though there is a fee waiver, testers were stuck on the fact that they would have to pay for the expungement process. They also wanted more upfront clarification on age restrictions. (You need to be over 18 to have your record expunged.)

We also added a visual task designed to understand how testers viewed the homepage. First, they reviewed the page for a few moments. Then we asked them to turn off their laptop screens and draw what they remembered. This exercise helped us see what was and wasn't obvious to them on that screen. For example, a majority of testers said they were totally willing to share information about Expunge.io on social media. But on the homepage, social media icons weren't easy to find. Only one tester noticed them.

Another big issue we found was that Expunge.io used a link with the word "adult" in it, and it was flagged as potentially inappropriate. This was an important insight for us, because families, other nonprofits and their schools might be using similar parental control software.

Testers also thought there should be a more clear call to action at the end of the process, even if they weren't eligible to get their own record expunged or if they didn't have a record. For example, tester MJ thought "there should a page where it says something like "Congratulations, you don't have a record! Here's what you can do to help others…"

For our second test, we visited Fenger High School, where students struggled not just with digital literacy, but with basic reading skills. It was an important reminder for us, to not rely on written surveys or materials that require a high reading level.

Fenger did remind us about something we're doing right. CUTGroup often asks people questions that go beyond the usability of app itself. We ask about testers' daily lives, the way they interact with their peers and communities, and their attitudes about relevant topics. For this Expunge.io test, that approach taught us a lot about the app's potential.

For example:

- Expunge.io will be an important way to teach people about why expungement matters, whether or not they actually have a record. Only 12 out of 21 testers thought that a juvenile record affects one's adult life
- Expunge.io will be highly relevant to youths' personal networks. Although the youth we met at Fenger are not necessarily eligible for expungement yet (because they are under 18 years old), 86% of our testers knew someone who was arrested as a juvenile
- Because Expunge.io will be so relevant and teaches people something new, it will seem like an idea worth sharing. The majority of testers share "important" information online. Look-

ing for Justice (#5) said, "I share some things on social media because it's funny or very important to me and everyone else should be warned and up to date"

Later in the spring, we returned to Mikva's JJC for a design session, and tested the site's branding and voice more fully. Smart Chicago will be working with the Mikva Challenge Juvenile Justice Council to share these results, and we hope to implement changes that will make Expunge.io work for more people.

References

Paul, Linda. (2014, February 4.) WBEZ. Why Is It So Hard to Expunge Juvenile Records in Cook County? Retrieved from: http://www.wbez.org/news/why-it-so-hard-expunge-juvenile-records-cook-county-105257

Sonja Marziano proctoring an Expunge.io test at Fenger High School.

O'Neil, Daniel X. (2014, December 12.) *Smart Chicago, Expunge.io, and Ecosystem.* Retrieved from: http://www.smartchicagocollaborative.org/smart-chicago-expunge-io-and-ecosystem/

Knight Foundation. (n.d.) *Expunge.io.* Retrieved from: http://www.knightfoundation.org/grants/201448416/

#12 – Roll with Me

Roll with Me is a website that helps residents find accessible transit directions in Chicago, and our test had two components— a week-long remote test and an in-person discussion.

We first met Roll with Me's developer, Mohammad Ouyoun, while he was a student at the Starter School. For one of his Starter School projects, he and his team created a prototype of the Roll with Me app. We were impressed, and after Starter School, we contracted with him to finish the app as part of the Civic Works Project.

We decided to focus on testing with individuals who have difficulties with stairs, and Mohammad predicted this site could be useful for temporary needs, too — parents with strollers, or individuals using carts or luggage.

A few great connections solidified because of this test. First, we'd already been planning to work with ADA 25, an initiative of the Chicago Community Trust celebrating the 25th anniversary of the Americans with Disabilities Act. We'd originally approached the ADA 25 committee because, overall, CUTGroup needed more people with disabilities. When we began to recruit for Roll with Me, they helped connect us with a few valuable testers.

Some of our testers were people with low vision, some used wheelchairs, and others used canes. We also connected with interested testers who could use stairs, so we broadened our criteria.

Here's what we most wanted to know:
- How does Roll with Me work within a user's normal, day-to-day routine? Is it easy to use on different devices?

- How important are the transit alerts for users?
- How does Roll with Me compare to users' other methods of finding transit routes? What features are users interested in having?

The first part of the test was conducted remotely. We'd anticipated that people wouldn't be very invested in a remote test, but in fact, we got a ton of helpful responses. We never got any two-word answers, or the sense that people were "just trying to get through it." One woman was searching for a cross street in the Austin neighborhood of Chicago and ended up being directed to Texas. Although she was frustrated, she gave us a very thorough and useful response by giving us step-by-step details on the remote test.

Other things we learned: 71% of testers normally check directions on their mobile device, and 50% thought that it was simple, easy, or straightforward to do that. The majority of testers found Roll with Me's transit alerts to be very useful, but a few testers thought that they were distracting from the route information.

Ten out of 14 people were able to make it to our in-person session. We hosted this as an unstructured evening at our offices, where we ordered dinner and held an informal focus group. Although we weren't completely sure what this would yield, soon participants were building on each others' thoughts, sharing experiences, and offering suggestions to improve everything, right down to the colors on the app. For example, we discussed how it can be difficult for people with certain visual impairments to differentiate between particular color pairings.

It was supposed to be a 45-minute discussion, but we instead stayed for an hour and a half, because it was so fruitful. We talked about the app, but we also just talked about technology in general. The best part is, now we have testers with disabilities in our network, who continue to come out for other tests, and they're an invaluable part of the CUTGroup family. All developers would benefit from working with them.

References

Starter School (n.d.) Retrieved from http://www.starterschool.com/

Smart Chicago Collaborative. (n.d.) Civic Works Project. Retrieved from http://www.smartchicagocollaborative.org/work/special-initiatives/civicworks/

Smart Chicago Collaborative. (n.d.) Mohammad Ouyon. Retrieved from http://www.smartchicagocollaborative.org/people/consultants/current-consultants/mohammad-ouyon/

Roll With Me. (n.d.) Retrieved from http://www.rollwithmeapp.com/

Making sure everyone gets to the right place.

Examples

Afterword

by Sonja Marziano

I participated in my first CUTGroup test on a very cold evening in November, 2013, in the Grand Boulevard neighborhood on the South Side of Chicago. As people streamed into the Hall Branch of the Chicago Public Library wearing heavy winter coats, I was excited and not really sure what to expect. I am not a UX researcher, and this was my first experience conducting UX testing. It surprised me that 19 residents from all over the city would come to a library (or, for some, a neighborhood) that they had never been to before, to give feedback on a website about food inspection data.

Not only did they come, but they stayed. Most testers stayed longer than the 45-minute time slot we allotted for each session because they wanted to keep talking with us, laughing with each other about their experiences with food poisoning, and giving us recommendations and ideas for improvements to the site.

I learned from this session how much people want to help make technology better, and that people know best about how they use technology and how they want technology to work for them.

There are too few opportunities for people in Chicago to be heard, and CUTGroup is a great lesson in how and why we need to invite people to participate in technology.

After that first test, I never stopped working on CUTGroup. The feeling after a test is infectious and energizing, and talking with people adds an immeasurable value to the work that we do in technology.

Since that November evening, we have more than doubled the number of CUTGroup testers to over 1,200; expanded into Cook County; and conducted 18 additional tests. We continue to modify and expand the ways we work with testers. I say all of this, not because it's a lot of work, but because the work is never done.

The last test we cover in this book is Roll with Me, but since then we have tested ten additional websites and apps.

You can stay up to date on our findings through the CUTGroup section of the Smart Chicago website: http://www.smartchicagocollaborative.org/work/ecosystem/civic-user-testing-group/

As we add more tests, we are also thinking about how we can reach more testers, how we can improve testing methods, how we respond to what testers tell us, and how we can make CUTGroup an integral part of creating better technology.

Reaching More People

We have started testing more websites that reach more people than ever before, such as the Chicago Public Schools website, the Ventra mobile app, and the Cook County website.

The expansion of the CUTGroup to all of Cook County was an important step. We knew we would be doing tests, such as wireframe testing of the new county website, that were created for all County residents. As we grow in Cook County, we are learning how to recruit in these new areas. We are also recruiting in some of the same ways that we did when we first started, reaching out first to libraries and computer centers and then determining neighborhoods where we will need to do more direct outreach.

But while we continue outreach in Cook County, we also wanted to continue being inclusive and reaching new people in Chicago. We have testers in every ward of the city, but we identified seven neighborhoods for more direct outreach, mainly centered in Chicago's Southwest side, which had the lowest number of sign-ups.

Making CUTGroup Work

We asked testers who had never participated in a test, "How can we make CUTGroup work better for you?" As we thought of ways to make CUTGroup more accessible, we decided to create a way for people to sign up and receive notifications about new tests through

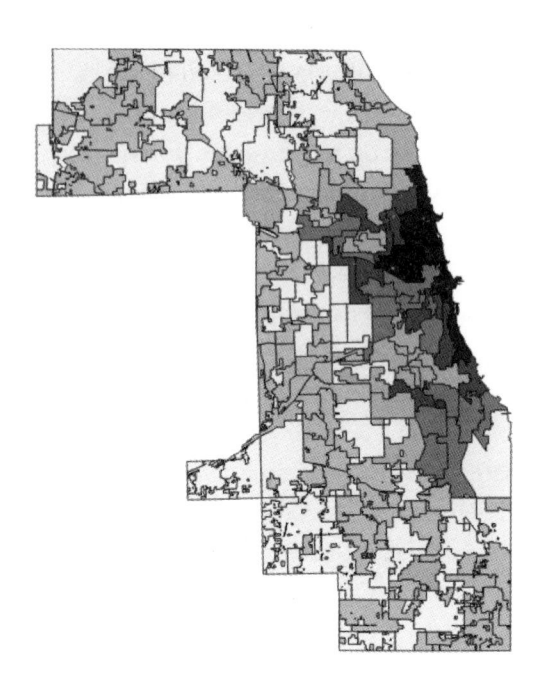

Map of CUTGroup testers, 2016.

text messaging. Approximately 29% of our testers say they connect to the internet primarily through public wifi or through a phone with a data plan. By adding text messaging, we serve residents who do not have regular access to the internet, making us more effective at reaching more residents and listening to their voices.

We also want to make it easier for developers or project managers to take the information they learn from CUTGroup and make changes to their websites and apps. At the same time, we know that the value of CUTGroup is helping residents' voices influence technology. To accomplish both of these goals, we gather more information early in the testing process that will drive feasible changes. Then after the test, we share those changes with our CUTGroup testers, to show them how they have real influence.

Everyone You Haven't Met Yet

Through the CUTGroup, we have built a community of people who want to make technology better. This is the most important piece of CUTGroup. If you want to incorporate UX testing into your work or create a CUTGroup in your city, my advice is simple: start with building community.

Invite everyone you know and then invite everyone you haven't met yet. Make it as easy as possible for anyone to join and participate in tests. Leave your office and visit a new neighborhood. Leave your assumptions aside (unexpected solutions can sometimes be the most valuable part of UX testing). Step back and listen to what your testers are describing. Keep asking questions. Provide space for your testers to tell you about what they would change, and then take their time and advice seriously. They are there to help you.

Then after the test is over, make real changes to your technology, and start it all over again.

Acknowledgements

Our last name is "Collaborative." It's how we live and how we work. With a team of just three people, there's really no other way. Here's a quasi-chronological list of the people who've helped us with the CUTGroup, and a short description of the huge contributions of each.

Julia Stasch, http://www.macfound.org/about/people/170/: Julia was a leader in the formation of Smart Chicago and has articulated many of the precepts of the CUTGroup for years. When writing the foreword to the 2007 report, "The City that Networks", http://goo.gl/yK2k4H, she said, "digital excellence is achieved when all who wish to can integrate the Internet comfortably into their lives — a state of active and meaningful participation that increases knowledge and enhances connections across time and place."

Terry Mazany, http://www.cct.org/about/our-staff/terry-mazany: Terry provides the framework for Smart Chicago to exist and oversees all our activities. His commitment to inclusion imbues all of our work, as it emanates from the Chicago Community Trust. His early encouragement of the CUTGroup was invaluable.

John Tolva, http://www.ascentstage.com/: John has been a supporter of the CUTGroup since the day it launched, on the inaugural meeting of Mayor Emanuel's Technology Industry Diversity Council.

Alaina Harkness, http://www.macfound.org/about/people/158/: Alaina has always encouraged Smart Chicago to try new things.

Mark Harris and Alya Adamany of the Illinois Science and Technology Coalition: http://www.istcoalition.org/. Our shared program, the Illinois Open Technology Challenge, http://illinoisopentech.org/ directly led to the creation of the CUTGroup.

Veronica Ludwig, http://about.me/veronicaludwig: Our conversations with Veronica on how to manage the Illinois Open Technology Challenge ("inviting these communities and citizens to the expo as a form of market research on existing state-specific civic applications") were formative.

Scott Robbin, http://srobbin.com/: Scott's website, Open Chicago ("Creating points of contact between developers and government"), http://www.openchicago.org/, launched in February 2012, and was an intellectual and technical precursor for our work. The idea that there should be a fluid relationship between developers and aldermen was just one conceptual step removed from the fluidity we now seek between residents and developers.

Chris Gansen, http://www.chrisgansen.com/. While he served as the first product manager for Smart Chicago, Chris was essential in the creation of the CUTGroup. He made the website, implemented our segmenting code, and helped design the entire system.

The Chicago Tribune Apps Team, http://blog.apps.chicagotribune.com/: The development team of Christopher Groskopf (https://twitter.com/onyxfish), Ryan Nagle (https://twitter.com/ryannagle), and Ryan Mark (https://twitter.com/ryan-mark), created the Chicago Boundary Service, http://boundaries.tribapps.com/, which is super-helpful in letting potential testers know what ward they live in.

Kyla Williams: As the second employee of Smart Chicago, Kyla is critical to all of our work. Her years of service in nonprofits and an ease with all people makes her an invaluable voice.

Randall Walker: Randall was an important part of the Smart Chicago team when we created the CUTGroup, and he was responsible for gift card distribution early on.

Emily Escarra: Emily was an important part of the Smart Chicago team when we conducted our first tests. She was a proctor at our first in-person test and wrote the analysis for our first test, FreedomPop.

Jason Kunesh, http://jdkunesh.com/: Jason was essential in sharing his experience in sensitive UX test design as we were first planning the CUTGroup.

Adam Steele, http://www.cdm.depaul.edu/people/pages/facultyinfo.aspx?fid=148. Adam lent advice on sound UX testing very early on in the development of our process.

Julie Harpring, http://carmentastreet.com/: Julie was a critical link in the creation of the CUTGroup. She designed and was the main proctor of our first in-person test, at the Kelly Library in the Englewood neighborhood, effectively giving us a crash course in the discipline of UX design. She also helped draft our tester release form.

Christopher Whitaker, http://civicwhitaker.com/: Christopher was critical in spreading the word about the CUTGroup in the Chicago civic hacker community. His steady output for Smart Chicago is always a huge asset to us. He also helped conduct a number of tests and composed the results of the Chicago Health Atlas test.

Jeff Murray of Chicago Nightlife Magazine: https://www.facebook.com/pages/Nightlife-Magazinenet/59605221592. Jeff helped us with very effective email marketing early on in the project.

Bryan Thompson: Bryan did much of the legwork to get the word out into communities so that we could have representation from the entire city.

Tom Kompare, http://about.me/tomkompare: Tom was the first developer to sign up for a CUTGroup test. His genuine curiosity about people and his earnest desire to make things better makes him the model partner for this endeavor.

Melissa Harris, https://twitter.com/ChiConfidential: Melissa attended our second in-person test and wrote a great column about it in the Chicago Tribune. This led to more interest in the model and more testers.

Brian Bannon, Commissioner, and the people of the Chicago Public Library: http://www.cityofchicago.org/city/en/depts/cpl/auto_generated/cpl_leadership.html and http://www.chipublib.org/. We're especially grateful to the leadership and staff at the libraries where we've conducted tests: Kelly Library in Englewood, Uptown Library, the Hall Library in the Grand Boulevard neighborhood, the Rogers Park branch, the Mayfair branch, the Clearing branch, and the Blackstone library in Kenwood. The public library is an essential resource for our work.

Theresa Bradley, https://twitter.com/tbradley. Theresa took an early interest in the CUTGroup. Conversations with her helped solidify methodology.

Erie Family Health Center, http://www.eriefamilyhealth.org/, for hosting our test on the Chicago Health Atlas.

Brenna Berman, http://www.cityofchicago.org/city/en/depts/doit/auto_generated/doit_leadership.html: As a member of our Advisory Committee, Brenna pushes us into thinking about how to use this model for civic engagement in novel ways.

Sonja Marziano: Sonja is the third employee of Smart Chicago and began running the CUTGroup in the Spring of 2014. She developed the system we use for analysis as well as for publishing raw test results. She runs the entire process— meeting with the developers, designing tests, and administering test day like nobody's business.

Vivelo Hoy, http://www.vivelohoy.com/: Managing Editor **Fernando Diaz** (@thefuturewasnow) brought the best possible spirit of collaboration with community members in our test at the Hall Library for EatSafe.co.

Marc Hebert, https://twitter.com/anthromarc: Marc attended the EatSafe.co test and gave us some great advice on anthropology-based design.

Ian Dees, https://twitter.com/iandees: Ian proctored the test for Open Street Map, where he is a member of the US Board.

Comcast NBC Universal: http://chicago.everyblock.com/: **Matt Summy** (https://www.linkedin.com/in/matthewsummy) and **Paul Wright** (https://www.linkedin.com/in/pewright) signed up for and helped proctor a test on their iPhone app, shortly after the relaunch of their website.

Waitbot, http://waitbot.com/: **Dave Turner** (https://www.linkedin.com/pub/david-turner/4/730/875) and friends signed up for and helped proctor a test on their wait-time app.

Two developers of Foodborne Chicago, Joe Olson, https://twitter.com/JOlson7168, and **Cory Nissen**, https://twitter.com/corynissen took a great personal in the Foodborne test and followed up with technology changes and analytics.

The John S. and James L. Knight Foundation: http://www.knightfoundation.org/. Knight was the first foundation to explicitly fund a CUTGroup test, for Foodborne Chicago, and has continued to support the CUTGroup program through the Knight Deep Dive project: http://www.smartchicagocollaborative.org/work/special-initiatives/deep-dive/.

The Chattanooga Code for America team, http://chitchatt.org/: **Jason Denizac** (https://twitter.com/_jden), **Jeremia Kimelman** (https://twitter.com/jeremiak), and **Giselle Sperber** (https://twitter.com/_giselles) were the first team to set up a CUTGroup outside of Chicago (http://tester.openchattanooga.com/).

Open Oakland, http://openoakland.org/: **Steve Spiker** and others in the open-gov movement there are implementing a user-testing group to meet their needs (https://github.com/openoakland/cutgroup). Correspondence with the learned **Andrea Moed** (https://www.linkedin.com/in/amoeda) was particularly helpful in the organization of this book.

The Build it! Bronzeville team: Joshua Engel, Ronnie Harris, and Jennifer Reinhardt, who won first place at the Center for Neighborhood Technology 2013 Urban Sustainability Apps Hackathon competition and proctored the test of their game app.

Chris Rudd who led the Mikva Challenge Juvenile Justice Council (JJC) and requested testing of Expunge.io with a part of his grant from the Knight Foundation. Chris was key in building, testing, and listening to youth voice to incorporate feedback into Expunge.io.

Mohammad Ouyoun who started as a Smart Chicago intern in July 2014. We first met Mohammad while he was a student at the Starter School. For one of his Starter School projects, he and his team had created a prototype of the Roll with Me app. We tested the Roll with Me app in early 2015.

Lindsay Muscato, http://lindsaymuscato.com/: Lindsay organized, edited, and produced this book.

Jason Harvey, http://jasonharveydesign.com/: Jason designed this book.

Most of all, thank you to the more than 1,400 Cook County residents who make up the CUTGroup. You're it.

About

Smart Chicago
The CUTGroup is a project of the Smart Chicago Collaborative, a civic organization devoted to improving lives in Chicago through technology. We work on increasing access to the Internet, improving skills for using Internet, and developing meaningful products from data that measurably contribute to the quality of life of residents in our region and beyond.

The Author
Daniel X. O'Neil is the Executive Director of the Smart Chicago Collaborative, a civic organization devoted to improving lives in Chicago through technology. He's helped make civic apps since 2000, including Killer on the Loose, CTA Alerts, and City Payments. He's designed and delivered digital skills training on lightweight tools since 2003. He was a co-founder of EveryBlock, the innovative microlocal news website, in 2007. He's written and published four books of poetry, written and directed three plays, and keeps four journals in real notebooks at any given time. More here: http://www.derivativeworks.com/, here: http://www.danxoneil.com, and here: https://twitter.com/danxoneil.

All images in this book are by Daniel X. O'Neil. Nearly 40,000 hi-res images licensed as Creative Commons: https://www.flickr.com/photos/juggernautco/.